A TEXT BOOK OF

PLANT ANATOMY (PAPER-VII)

&

PLANT METABOLISM (PAPER-VIII)

FOR

B.Sc. Part II, (Botany) DSC-D13 & DSC-D14
(Semester - IV)

As Per New Revised CBCS Syllabus of Shivaji University, Kolhapur
With effect from June 2018

Dr. S. K. Khade
M. Sc., Ph. D.
Member of faculty in Science and Technology
BOS (Botany), S. U. Kolhapur
Former BOS and Senate Member, S.U., Kolhapur
Department of Botany, Padmabhushan
Vasantdada Patil Mahavidyalaya, Tasgaon.

Prof. (Dr.) N. A. Kulkarni
M. Sc., Ph. D.
Professor and Head Department of
Botany,
Padmabhushan Vasantdada Patil
Mahavidyalaya
Tasgaon.

Dr. V. B. Shimpale
M. Sc., Ph. D., FIAAT
Member of BOS (Botany), S. U. Kolhapur
Department of Botany, The New College, Kolhapur

Dr. G. G. Potdar
M. Sc., Ph. D., FIAAT
Department of Botany,
Yashwantrao Chavan College of Science,
Karad

B.Sc. Part II (BOTANY) PAPER VII & VIII   **ISBN: 978-93-89944-29-7**

First Edition : **February 2020**

© : **Authors**

Published By:

NIRALI PRAKASHAN

Abhyudaya Pragati, 1312, Shivaji Nagar
Off J.M. Road, PUNE – 411005
Tel - (020) 25512336/37/39, Fax - (020) 25511379
Email : niralipune@pragationline.com

➤ DISTRIBUTION CENTRES

PUNE

Nirali Prakashan : 119, Budhwar Peth, Jogeshwari Mandir Lane,
Pune 411002, Maharashtra. Tel : (020) 2445 2044,
Fax : (020) 2445 1538, Email: bookorder@pragationline.com,
niralilocal@pragationline.com

Nirali Prakashan : S. No. 28/27, Dhyari, Near Pari Company, Pune 411041
Tel : (020) 24690204 Fax : (020) 24690316
Email : dhyari@pragationline.com,
bookorder@pragationline.com

MUMBAI

Nirali Prakashan : 385, S.V.P. Road, Rasdhara Co-op. Hsg. Society Ltd.,
Girgaum, Mumbai 400004, Maharashtra
Tel : (022) 2385 6339 / 2386 9976, Fax : (022) 2386 9976
Email : niralimumbai@pragationline.com

➤ DISTRIBUTION BRANCHES

JALGAON

Nirali Prakashan : 34, V. V. Golani Market, Navi Peth, Jalgaon 425001,
Maharashtra, Tel : (0257) 222 0395,
Mob : 94234 91860

KOLHAPUR

Nirali Prakashan : New Mahadvar Road, Kedar Plaza, 1st Floor Opp. IDBI Bank
Kolhapur 416 012, Maharashtra. Mob : 9850046155

NAGPUR

Nirali Prakashan : Above Maratha Mandir, Shop No. 3, First Floor,
Rani Jhanshi Square, Sitabuldi,
Nagpur 440012, Maharashtra Tel : (0712) 254 7129

DELHI

Nirali Prakashan : 4593/15, Basement, Aggarwal Lane, Ansari Road,
Daryaganj, Near Times of India Building, New Delhi 110002
Mob : 08505972553

BANGALURU

Nirali Prakashan : Maitri Ground Floor, Jaya Apartments, No. 99, 6th Cross,
6th Main, Malleswaram, Bangaluru 560 003, Karnataka
Mob : +91 9449043034
Email: niralibangalore@pragationline.com

www.pragationline.com   **info@pragationline.com**

PREFACE

This is the sixth book in sequence after the successful writing of the books for new B.Sc. I and B.Sc. II (Semester - III) CBCS syllabus along with two practical handbooks for B.Sc. I and B.Sc. II students of Shivaji University, Kolhapur. First of all thanks for the response by the faculty and students.

The present book is intended for the use of B.Sc.II Botany students of Shivaji University, Kolhapur. As we have made all the attempts to make the book easy and simple and at the same time to upgrade the subject knowledge of the students.

In this book the efforts are made to collect and present maximum information with respect to various aspects of botany like Plant Anatomy, Organization of Plant Body, Tissue Systems, Enzymes, Nitrogen Metabolism, Respiration, Seed Dormancy and Germination. Attempts are made to make the topic simple with the help of neat and labeled diagrams.

The authors take the opportunity of thanking to Prin. (Dr.) Milind Hujare, Padmabhushan Dr. Vasantraodada Patil Mahavidyalaya, Tasgaon. Prin. (Dr.) V. M. Patil, The New College, Kolhapur and Prin. (Dr.) S. B. Kengar, Yashvantrao Chavan College of Science, Karad for the kind support and inspiration to write this book.

We are thankful to Nirali Publication, Pune for making us a part of their team of authors. We are thankful to Mr. Dineshbhai Furia and Jignesh Furia for publishing this book. We also thanks to Mr. Girish Redkar (Head, Marketing Department), for their co-operation in publishing this book.

We are also thankful to Mr. Virdhaval Shinde (Marketing Executive, Kolhapur District) and Mr. Ashok Nanavare (Marketing Executive, Sangli District) for their kind co-operation. We are very much thankful to Ms. Roshan Khan and Ms. Prachi Sawant for a neat and error free printing of this book.

We hope that this book will be found useful to the students and teachers of life sciences. We will be happy to receive constructive suggestions from the readers to improve our efforts in the future.

— **Authors**

SYLLABUS

(SHIVAJI UNIVERSITY)
PAPER VII : PLANT ANATOMY

1. ORGANIZATION OF HIGHER PLANT BODY AND TISSUES
(22)

1.a Organization of higher plant body (10)

1.1 The Plant organs

1.2 Development of plant body

1.3 Internal organization

1.b Meristematic and Permanent Tissue (12)

1.4 Meristem:

    a) Introduction, Characteristics and Classification of meristems based on position

    b) Theories of structural development - i) Apical cell theory ii) Histogen theory iii) Tunica Corpus theory.

1.5 Permanent tissue: i) Simple tissue- Parenchyma, Collenchyma and Sclerenchyma, ii) Complex tissue: Xylem and Phloem

1.6 Types of Vascular bundles

2 PRIMARY AND SECONDARY STRUCTURE OF PLANT BODY AND TISSUE SYSTEMS
(23)

2.a: Primary and secondary structure of plant body (12)

2.1 Primary structure of Monocotyledon and Dicotyledon root, stem and leaf.

2.2 Normal secondary growth in Dicotyledon root and stem.

2.3 Anomalous secondary growth in Bignonia (Dicot) and Dracaena (Monocot) stem.

2.4 Periderm and Lenticel

2.b: Tissue systems (11)

2.5 Epidermal tissue system

2.6 Secretary tissue system

2.7 Mechanical tissue system

PAPER VIII : PLANT METABOLISM

1. ENZYMES AND NITROGEN METABOLISM (22)

1.a Enzymes (12)

1.1 Introduction

1.2 Classification and Nomenclature of enzymes

1.3 Structure and properties of enzymes

1.4 Mechanism of enzyme action- Lock and Key hypothesis and Induced fit hypothesis.

1.5 Factors affecting enzyme activity temperature and pH.

1.6 Enzyme inhibition

1.b Nitrogen Metabolism (10)

1.7 Introduction

1.8 Biological Nitrogen Fixation- Asymbiotic and Symbiotic

1.9 Mechanism of Nitrogen Fixation

1.10 Nitrate reduction

1.11 Ammonia assimilation

1.12 Nif genes

2. RESPIRATION, SEED DORMANCY AND GERMINATION (23)

2.a Respiration (12)

2.1 Introduction

2.2 Types of respiration

2.3 Glycolysis

2.4 Formation of Acetyl Co A

2.5 TCA cycle

2.6 ETS in mitochondria

2.7 Fermentation

2.b Seed Dormancy and Germination (11)

2.8 Concept of dormancy

2.9 Causes of dormancy

2.10 Methods of breaking of seed dormancy.

2.11 Seed germination- Introduction and types (Epigeal, Hypogeal and Viviparous).

2.12 Factors affecting seed germination

2.13 Biochemical changes during seed germination.

❖❖❖

CONTENTS

1
CHAPTER

ORGANIZATION OF HIGHER PLANT BODY AND TISSUES

1.1 INTRODUCTION

'Anatomy' is a branch of morphology that deals with the structure of organisms. Plant anatomy is dealing with the internal structure and their organization. Basically the word *'anatomy'* is derived from two Greek words, *'ana'* meaning as under or separated into pieces and *'temnein'* means to cut, hence it is a study which is done after sectioning of the plant parts. So, anatomy can be defined as, 'it is a branch of botany that deals with the internal structure and organization of the plant parts'.

Marcelli Malpighi (1628-1694) and Nehemiah Grew (1641-1712) are the two pioneer workers who worked on plant anatomy. Grew an English physician and botanist worked on plant tissues and Malpighi an Italian physician worked on the various tissues of vascular plants and have published their work in sixteenth century.

The 'Plant Kingdom' shows great diversity in their forms and organizations. It ranges from unicellular microscopic plant (algae) to the giant multicellular trees (angiosperms). When plant is composed of complex tissues especially with vascular tissues then it is referred as higher plant and due to presence of vascular tissues the higher plants also called as vascular plants. It is the most evolved group of plants which constitute the dominant part of earth's vegetation. It also shows the considerable variation in the tissues and organs of the plant body.

The plant body is typically consists of two parts i.e. root and shoot. Generally the shoot grows above the ground, upward into the air and comprises of stem, it bears leaves at the nodal region while the root is generally subterranean and forms underground plant organs and grows downward into the soil.

A mature seed after germination produces a plumule and radical, the plumule gets modified into the aerial shoot system and radical forms an underground root system. After development of shoot and root apices it produces number of specialized cells and these specialized cells carry various physiological and mechanical functions in the plant body. The phenomenon of formation of specialized cells is referred as cell differentiation while the process of formation of tissue and organ is referred as development or morphogenesis (morpho = form; genesis = origin). The morphogenesis is responsible for formation of different organs in the plant body. Thus the process of growth and the differentiation both are responsible for formation of different tissues, organs and the body of an organism. During the formation of plant body the genetic factors along with environmental factors plays important role.

1.2 THE PLANT ORGANS

Arnold (1947) have already pointed out that there is differentiation in the organization of higher plant quoting the Psilophytes as an example. Basically the plant body is originated from the morphologically unicellular zygote, after development in the zygote it produces the embryos and then embryo get converted into the mature plant. It is observed that the higher plant is originated by the evolutionary specialization through a long period.

This specialization is not only responsible for the establishment of morphological and physiological differences between the various parts of the plant body but also responsible for the development of the concept of plant organs.

Although there is a great diversity in the size, structure and forms of the higher plants there is a fundamental structural uniformity. The structural uniformity is that each higher plant has an axis. The lower part of this axis which is subterranean is known as root while the aerial part is known as stem. Stem and root bears appendages of four types (Fig. 1.1), these are central axis, leaves, emergences and hairs.

1. **Central axis:** A central axis of a typical seed plant basically consists of two parts, the shoot axis and root axis. The characteristic feature of shoot, it grows in upward direction above the ground and produces stem, the stem is characterized by the presence of leaves on them. The primary functions of the stem are: (i) it bears leaves, (ii) it is associated with the conduction of water and minerals from root to the leaves, (iii) also perform conduction of food material from leaves to various parts of the plant and (iv) production of reproductive structures (the flower). The root is another part of the central axis, the root axis mostly grows in opposite direction of the shoot and grows deep in the soil. The main function of root is anchorage and absorption of water and minerals from the soil.

2. **Leaves:** It is the most important organ in which vascular strands passes from the stem. Leaves are only present on the stem and do not occurs on the root. Leaves have a definite order of development and are in close association with the stem. Thus it is stated that the leaves are nothing but lateral expansions of the stem. After maturity flowers are produced on the aerial axis, it is the reproductive organ of a plant which is specially meant for sexual reproduction. It is an accepted view that the flower is homologous to a shoot and floral parts with the leaves. Mostly a flower is composed of sepals, petals, stamens, carpels and sometimes also sterile members.

   The axillary buds are present in the axils of the leaves. These axillary buds are responsible for production of vegetative branches or/and reproductive branches. Such axillary branches also develop in the root system, but not from axillary buds. The root branches are endogenous in origin and develop from interior region like the pericycle. These root branches have connections with conducting strands of the central axis of the root. The main root produces primary, secondary, tertiary branches which increase the surface area of the roots and thus helps to absorb the water and minerals from the soil.

3. **Emergences:** These are the second type appendages, mostly originated from the external portions of the stem. These are originated from the cells of epidermis and cortex. The presence of prickles on the stem of rose is a well known example of emergences.

4. **Hairs:** Hairs composes outermost layers of the stem and leaves forming appendages of the third rank. These are superficial out growths.

Thus, the body of higher plants is consists of root, stem, leaves and epidermal outgrowths. Each organ has its own morphological structures (peculiarity) and specific functions. Also they have their specific tissue system to perform their functions.

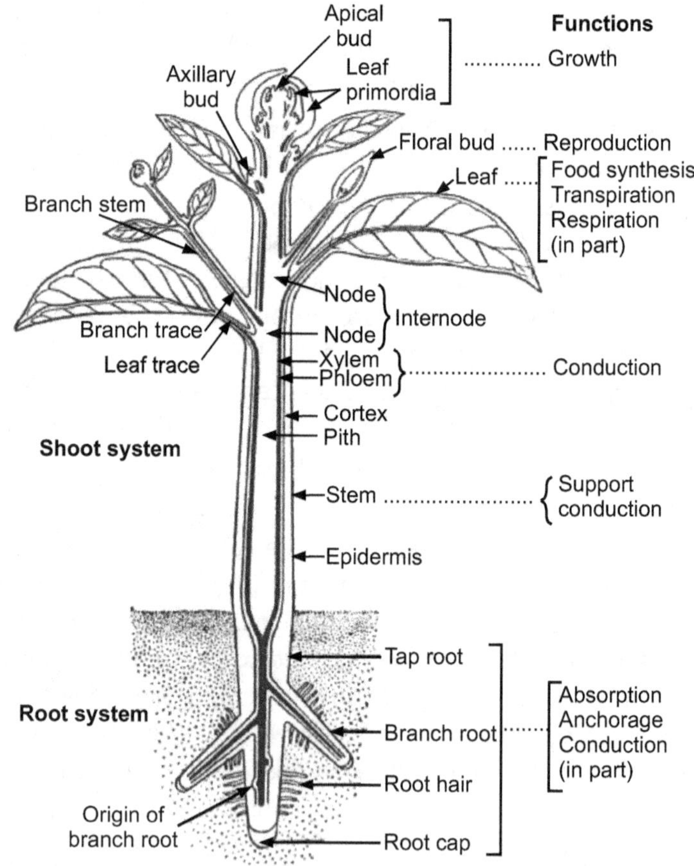

Fig. 1.1: Plant Body showing fundamental parts

1.3 DEVELOPMENT OF PLANT BODY

Higher plant produces the seeds and after seed germination it starts its life. The mature plant produces flowers, which get fertilized and produce seeds. A seed is composed of embryo and endosperm, after getting the favourable conditions the seeds starts to germinate and produces a new plant body (Fig. 1.2a). Basically after fertilization zygote is formed which is unicellular, after further development it develops into embryo. Hence, a vascular plant begins its existence as a morphologically simple unicellular zygote (2n). As zygote do not undergo meiotic division it produces the sporophytic plant body, so all higher plants shows sporophytic plant bodies. The development of the sporophyte involves divisions and differentiation of cells and also involves in organization of cells into the tissues and tissue system.

In comparison with sporophyte the embryo of the seed plant possesses relatively simple structure and it is only an axis which bears one or more cotyledons. The tissues that are present in the embryonic axis are meristematic in nature (meristematic tissues give rise to various organs and are responsible for growth of the plant). After further development the embryonic axis produces two distinct regions known as a plumule and radicle. At the apex plumule have zone of meristematic tissues known as shoot apical meristem (SAM) while meristematic tissues that are present at root apex produces root apical meristem (RAM). The plumule is responsible for formation of shoot organs like stem, leaves and flowers while radicle produces root organs.

(A) Shoot Apical Meristem (SAM): In young plants it is located at tip of the stem, it is a dome shaped structure that is surrounded by curved leaf primordia. The SAM perform two important functions like (i) it forms primordia which gives rise to lateral organs and (ii) also to perpetuate itself (to remain in the form of original embryonic cells). The shoot apical meristem have a region that have dividing capacity and this meristematic region produces all aerial organs of the plant body including stem, leaves, branches and flowers. During this development, SAM cell divides to form two celled structure, out of these two cells one cell remains as meristematic cell (it remain in the form of original embryonic cell) while other cell undergoes for elongation and differentiation. Generally SAM is composed of two zones, the outer zone is tunica and the central zone is corpus (Fig. 1.2b).

The Tunica layer generally composed of one to several layers of cells, the cells of tunica region divides anticlinally (the plane of division is right angle to the surface of the plant) or perpendicular to the main axis of the meristem. The tunica region is responsible to produce the outer layers of the plant body like epidermis of the plant body. While the corpus tissues divides more randomly in both anticlinal (perpendicular) and periclinal planes (parallel to the tissues) and produces bulk of tissues of the internal tissues of the stem and leaves. After the development of SAM tissues is produces nodes on the stem and from these nodes leaves are originated. All this process is controlled by specific genes along with hormonal action. After attaining certain vegetative growth it produces floral organs on the shoot axis.

(B) Root Apical Meristem (RAM): In comparison with SAM, RAM is structurally simple (Fig. 1.2c) because it is not responsible for the formation of branches in the root. In root system the branches are produced endogenously from pericycle region which is a surrounding region of vascular zone. The root tip is covered by root-cap as root has to penetrate into the soil particles and sometimes soil particles may be hard in nature. In addition with protective function, root cap also secretes mucilaginous substances called mucigel. This mucigel provides lubrication to the root-tip when it moves forward in the hard soil. Structurally the root apical meristem (RAM) is present just below the root cap. This zone is divided into two zones i.e. zone of slowly dividing cells called quiescent zone and zone of fast dividing cells. The cells of the RAM are responsible for elongation of roots, formation of new tissues in the root and also helps in regeneration of the rootcap. Root apical meristem also responsible for the formation of other different tissues that are present in the root like pericle, pith etc.

Primary and secondary growth: As mentioned above, the primary growth is that which takes place by primary tissues. The tissues that have embryonic origin are known as primary tissues and the growth taking place by such tissues is referred as primary growth, ultimately the tissues that are first formed are known as primary tissues. For example the xylem which is formed by cells that are originated from embryonic axis is referred as primary xylem like wise primary phloem and other tissues. Generally in most of vascular cryptogams and monocotyledons, the plant is composed of most of the primary tissues. While in higher plants like gymnosperms and most of the dicotyledons and few members of monocotyledons there is an increase in thickness of stem and root which is due to secondary growth.

Basically few tissues that undergoes for differentiation in course of plant growth, out of these tissues few tissues regain their cell division capacity and such tissues are referred to as secondary tissues and the growth taking place by such tissues is referred as secondary growth. Therefore the tissues formed as the result of secondary growth are called as secondary tissues. Generally during the secondary growth new types of cells are not formed. But the girth of plant increases due to secondary growth and also secondary tissues are protective in nature. Thus, a secondary body composed of secondary tissues is added to the primary body which is composed of primary tissues. A special type meristem is present in the plant known as 'cambium' and it is the most common tissue which is responsible for secondary growth. The cambium arises between the primary xylem and the primary phloem and produces new xylem towards the pith region and phloem at the epidermal region. Therefore the secondary xylem and phloem is found entirely within the central cylinder and mostly in between the primary xylem and phloem patches. In addition with cambium, a cork cambium or

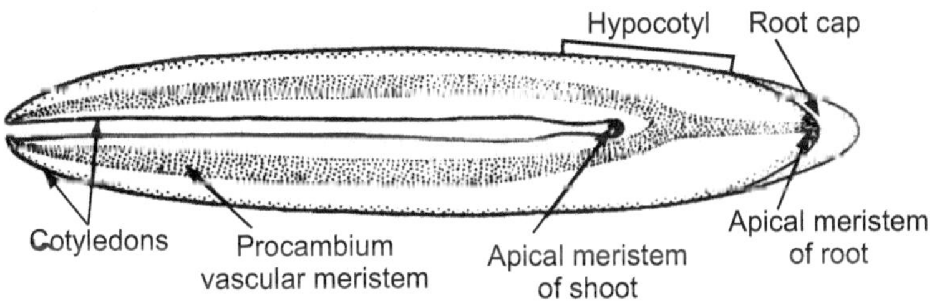

Fig. 1.2 (a): Mature embryo of seed plant

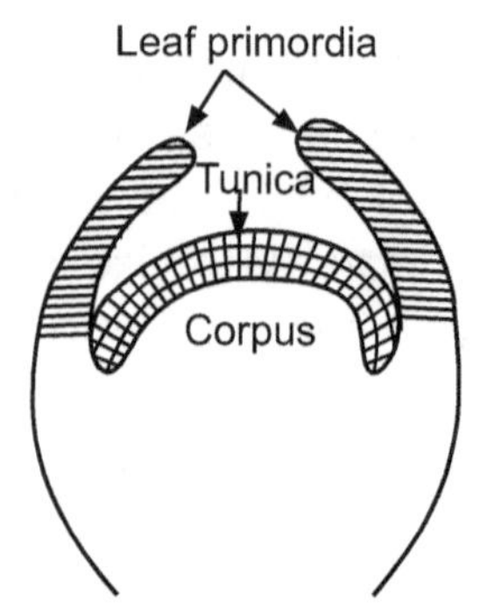

1.2 (b): Two zones of shoot tip (SAM) region

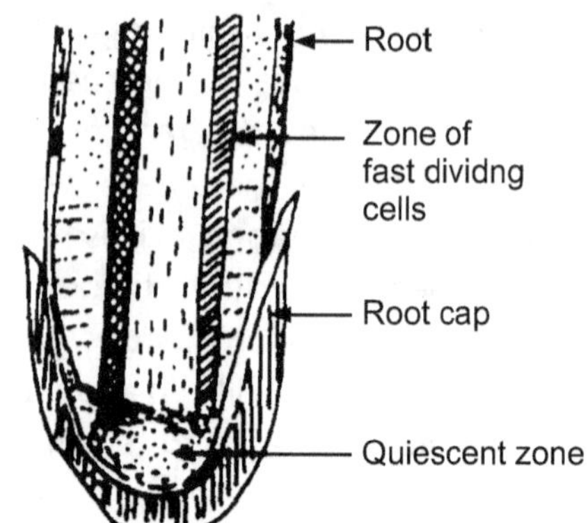

1.2 (c): Two zones in Root tip (RAM) region

phellogen is present at the peripheral region of the axis and produces a periderm. This periderm is also a secondary tissue which is mostly associated with protection of internal tissues especially when the primary epidermal layer is disrupted during the secondary growth.

1.4 INTERNAL ORGANIZATION

Body of higher plants is composed of different types of tissues and these tissues are distributed in the various regions of the plant body. The morphologic units (cells) of the plant body are associated in various ways with each other. The plant organs are composed of several kinds of tissues. The larger units of tissues may show topographic continuity or physiologic similarity or both together and a system formed by such units is called as a tissue system. Thus the plant body shows complex structure formed by different cells, the cells shows variation in their functions and in the combination of other cells. According to Sachs (1857), the plant body of a vascular plant is composed of three zones i.e. (i) the epidermal (ii) the vascular (iii) the fundamental or ground zone. The separation of these zones is based on the position and the basic functions of the tissues. The partly differentiated derivatives from the meristem can be classified as –protoderm, ground meristem and procambium. These meristematic tissues are responsible to make dermal, fundamental (ground) and vascular system, respectively.

Promeristem

↓

Protoderm	Ground meristem	Procambium
↓	↓	↓
Zone of epidermal cell	Zone of ground tissues	Zone of vascular elements

In general the anatomy of plant body there are three vegetative organs like stem, root and leaf. These are distinguished by the relative distribution of the vascular and the ground tissues. In stem the epidermal tissues are present on outer side of the plant organs, the cortical zone (ground tissues) are present in between the epidermis and vascular region while vascular tissues are found between the epidermis and the central axis.

In the root, the pith may be absent and cortex is generally shade during secondary growth. The primary vascular tissues (primary xylem and primary phloem) are commonly being arranged in the form of ring of bundles like in cross section of stem. But during the secondary growth the original primary vascular tissues may be obscured by secondary vascular tissues.

Many interconnected strands (bundles) are found in the ground tissues of the leaf. Generally in leaf the ground tissues are composed of photosynthetic parenchyma (chlorenchyma) also known as mesophyll.

The common tissues that are present in the various zones are:

(A) Epidermal zone: It is the outermost zone of the plant body. It present from apex of the shoot to the root. Generally it is composed of single layered but sometimes it may be multilayered. The epidermal tissues are associated with the protection of internal tissues from the external environment and also to check the rate of transpiration.

(B) Cortical zone: It is the second zone of the internal organization of the plant body. The tissues that are present in this zone are of various types and it is multilayered zone. The cortical zone represents the ground or fundamental tissue system of the plant body. Commonly this zone is composed of parenchymatous

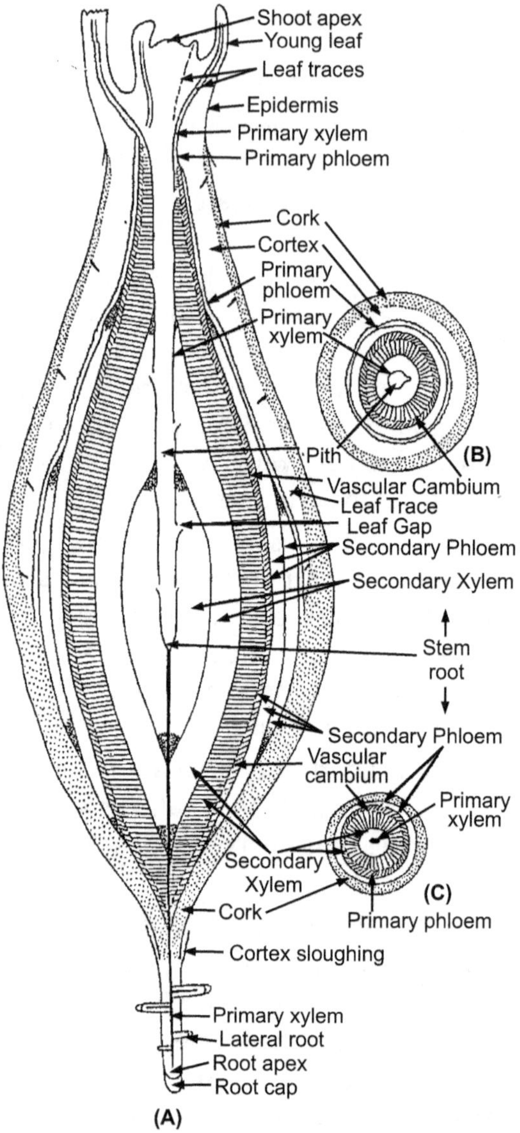

Fig. 1.3: Internal organization of plant body

tissue. These tissues are concerned with photosynthesis, storage, wound healing and some other specialized functions like arenchyma, these are associated with floating of the aquatic plant body. In this zone collenchyma tissues are also present which gives mechanical strength or physical support to the herbaceous plant body.

(C) Vascular zone: It is the third innermost zone in the plant body, it is surrounded by cortex and composed of vascular tissues and hence also referred as vascular strand or stele. As it is associated with the conduction of water, mineral and solutes, it is also referred as conducting strand or conducting cylinder. The vascular cylinder is composed of two important tissues i.e. xylem and phloem, by combination of these tissues plant body composes the complex body. The xylem tissues are concerned with water conduction, storage and support, the principle conducting cells are tracheids and vessels. While phloem tissues are mainly associated with translocation of solutes and storage of food. The principle conducting cells that are present in the phloem are sieve cells and sieve tubes. Mostly in roots the cortex and vascular cylinder get separated by the presence of endodermis. The endodermis prevents water conducting cells from becoming clogged with air and acts as an air drum. These tissues are also connected with maintenance of root pressure and sometimes also serve as protective tissue. Physiologically the cells of endodermis regulate the absorption of water and other substances from the soil.

QUESTIONS

A. Multiple Choice Questions:

1. Anatomically higher plant body can be divided into .... zones.
   - (a) 2
   - (b) 3
   - (c) 4
   - (d) 5

2. .... is responsible for production of root hairs.
   - (a) Stele
   - (b) Pericycle
   - (c) Endodermis
   - (d) Epidermis

3. .... is responsible for production stem.
   - (a) Plumule
   - (b) Radical
   - (c) Vascular bundle
   - (d) Apical meristem

4. .... is responsible for production of root.
   - (a) Stele
   - (b) Pericycle
   - (c) Endodermis
   - (d) Padical

5. .... is tissues are always living in nature.
   - (a) Permanent
   - (b) Meristematic
   - (c) Endodermis
   - (d) Stelar

6. The cell that have cell division capacity is referred as .... cell.

   (a) epidermal   (b) meristematic

   (c) permanent   (d) cortex

7. Mostly .... possess secondary tissues.

   (a) monocot   (b) dicots

   (c) bryophytes   (d) pteridophytes

8. Generally cortical zone of higher plants is composed of .... tissues.

   (a) Stelar   (b) epidermal

   (c) Parenchyma   (d) xylem and phloem

9. In higher plant body conduction of water and solutes takes place by .... tissues.

   (a) Stelar   (b) Pericycle

   (c) Endodermis   (d) Epidermal

10. The phloem tissues is mainly associated with conduction of ....

   (a) water   (b) food material

   (c) solutes   (d) air

Answer key:

(1) – b, (2) – b, (3) – a, (4) – d, (5) – b, (6) – c, (7) – b, (8) – c, (9) – a, (10) – b.

B. Broad Questions:

1. Describe with suitable diagram organization of a higher plant body.

2. Describe in brief the internal organization of a higher plant body.

3. Explain with diagrams SAM and RAM in plants.

C. Write Short Notes on:

1. SAM

2. RAM

3. Primary growth in higher plants

4. Secondary growth in higher plants

5. Cortical zone in higher plants

6. Vascular zone in higher plants

2

CHAPTER

MERISTEMATIC AND PERMANENT TISSUES

2.1 INTRODUCTION

In lower plants, plant body is composed of either single cell or colonies of similar cells and mostly the morphological differences in the cells are not noticed in vegetative and reproductive parts. It is stated that during the evolution in the plant kingdom, the plant tissues get modified by the gradual complexity from the simple tissues. Hence the higher plants are composed of diverse types of cells that vary in their shape, size, origin and function. A group of cells which have more or less similar size, shape, same origin, same methods of development and same functions are known as tissues. This definition is quite ambiguous because in complex tissues like xylem and phloem, although these tissues are composed of distinctly dissimilar cells but finally forms a tissue (xylem and phloem are composed of different types of cells). The body of higher plants is composed of more complex tissues which can be differentiated into vegetative and reproductive tissues. The growth and development in the plant body mainly takes place by addition of new cells, this addition is due to the activities of the meristematic cells that are present in the specific regions of the plant body. During cell division, the meristematic tissues produce derivatives (which after some development become permanent tissues) and at the same time some meristematic cells perpetuate its cell division capacity (such cells are known as initiating cells) which continues their cell division capacity throughout the life cycle. In other words cells within meristem are self-renewing that means every time they divide, one new cell remains meristematic while the second cell to become a programmed or specialized mature cell. Due to this activity the size and height of plant and plant organs get increased throughout its

(2.1)

life. The tissues can be classified on the basis of different characters like nature of cells, origin and methods of development, position in the plant body etc. However, the tissues primarily can be divided into two main group i.e. (i) Meristematic tissues or meristems and (ii) Permanent tissues.

2.2 CHARACTERS OF MERISTEMS

Meristem tissues are also commonly known as *'meristems'*. The term *'meristem'* is composed of Latin word *'meristos'* means *divisible* and thus a meristem can be defined as an *'immature, not fully differentiated cells which possess the capacity of division'*. The meristematic cells have following important features (Fig. 2.1):

1. The cells are normally isodiametric or spherical or polygonal in shapes.

2. They are without intercellular spaces and compactly arranged.

3. They have thin, homogenous and cellulosic cell wall. Secondary wall is normally absent.

4. Each cell has a large distinct and prominent nucleus that associated with dense cytoplasm. Cytoplasm with or without vacuoles and mostly lack of ergastic substances (very rarely tannins may be present).

5. Plastids are in the form of proplastid stages (it is a small colourless organelle that gives rise to a plastid).

6. The most important feature of meristematic cells is that these cells have cell division capacity. In fact, meristems are formative region from where new cells are added to the plant body.

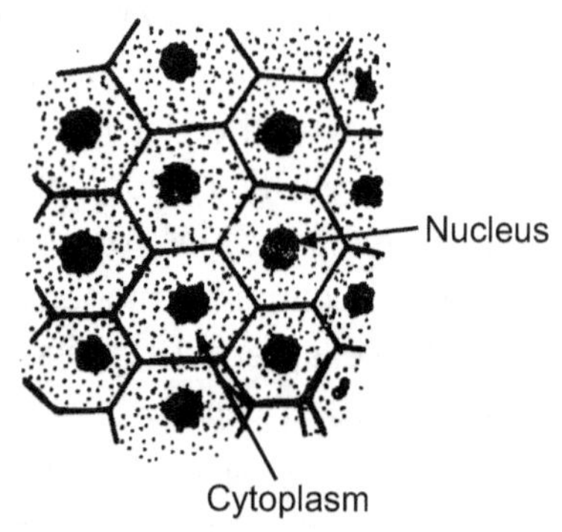

Fig. 2.1: Structure of Meristematic tissues

2.3 CLASSIFICATION OF MERISTEM BASED ON POSITION

Meristems are normally classified on basis of their origin, plane of division, functions and position in the plant body. But one of the most common groupings of plant meristems is based on their position in the plant body. This classification divides the meristematic tissues into the plant body on the basis of their location in the plant body. For example the apical meristems that is the meristems which located at the apices of main and lateral shoots and roots while lateral meristems means the meristem that arranged parallel with the sides of the organ in which they occur. Following are the types of meristems based on their position in the plant body, three categories are recognized (Fig. 2.2).

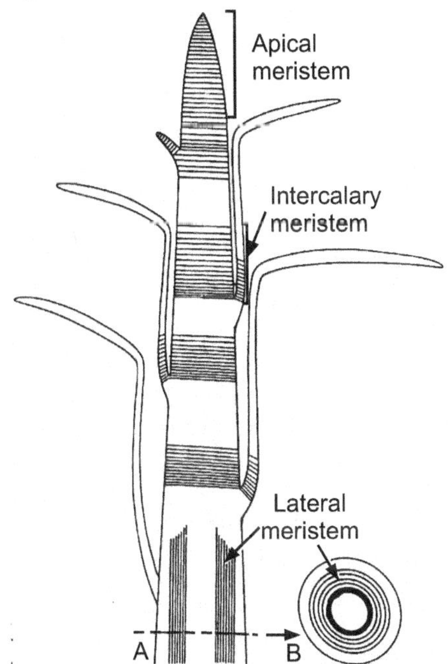

Fig. 2.2: Classification of meristem based on position

1. **Apical meristem:** These always occur at the apices of the main stems, roots and at the lateral branches of root and stem. As the length of plant body increases due the activity of apical meristem, it is also referred as growing points, these initiating points may be composed of single cell or group of cells. It includes both

promeristem as well as primary meristem. The initiating cells may be in groups or solitary. Mostly solitary apical cells are present in the pteridophytes while group of initiating cells occurs in all other vascular plants. These apical meristematic tissues divides and gradually undergoes for differentiation and produces the permanent tissues which forms the primary plant body of the plant.

2. Intercalary meristem: Basically these are cells of apical meristems which are separated from the apex during the growth of the axis, so these lies between the regions of permanent tissues and considered as a part of primary meristem. These meristems mostly found in the stem and leaf sheaths of many monocotyledons, especially in the grasses and in the horsetail (*Equisetum*). These meristems have very short life span, because these tissues either disappear soon or merge with surrounded permanent tissues. Like the apical meristem, intercalary meristem gives rise to the primary permanent tissues that forms the primary plant body.

3. Lateral meristem: Generally these meristems occur laterally to the axis and arranged parallel to the sides of organs and normally divide periclinally or radial and give rise to secondary permanent tissues. It is mostly present in the dicotyledons and gymnosperms. These are responsible to increase the thickness of the plant body. The cambium of the vascular bundle and phellogen (cork cambium) are common examples of lateral meristems. It should be noted that the cambium is an example of primary meristem while phellogen is the example of secondary meristem.

2.4 THEORIES OF STRUCTURAL DEVELOPMENT

Wolff (1759) first time studied the shoot apex in details and pointed out that the apices of the plant organs have undifferentiated growth apices, these apices may responsible for the elongation of the organs and plant body. Subsequently many workers have discussed the morphology and orientation of cells in the meristems, the number of meristematic tissues, the arrangement of meristematic tissues and also the plane of cell divisions in the meristematic tissues. So, several theories have been proposed by several authors to explain the mode of growth found in shoot apical meristem. All these theories are well supported by well examples. Some important theories are as follows:

i) Apical cell theory: This theory first time proposed by Hofmeister (1857) and supported by Nageli (1878). According to this theory, a solitary (single) apical cell is the structural and functional unit of apical meristem which is responsible for the apical growth in the plant body (Fig. 2.3). According to his view, like in the lower plants higher plants may prevails this single apical cell in the growing apices. However, subsequent workers have pointed out that such organization has been found only in cryptogams and not in phanerogams. So, this may be applicable only in cryptogams and not applicable for phanerogams (seed plants). The apical cell theory was superseded by the histogen theory.

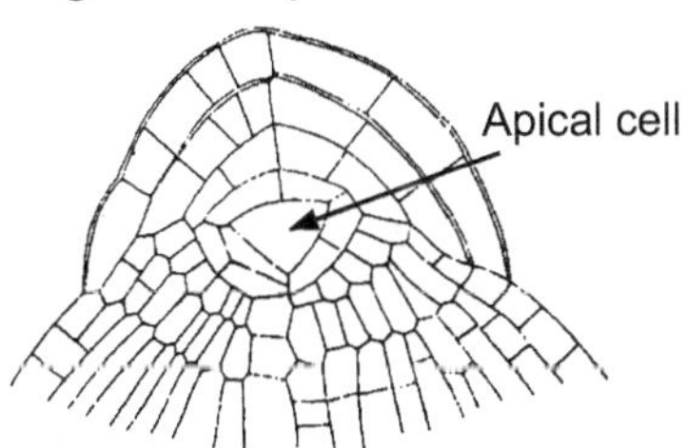

Fig. 2.3: Apical cell theory

ii) Histogen theory: This was proposed to make attempts towards interpretation of the growing apices of seed plants. The theory was proposed by Hanstein in 1870 and supported by Strasburger (1878), according to this theory the meristematic zones have group of initials (mass of meristematic tissues) instead of a single cell as proposed in the apical cell theory. The meristematic regions have three distinct zones or layers or histogens (histogens is a zone of clearly differentiated cells). According this theory the apical meristem or growing region of the stem and the root are composed of small mass of cells which are similar and have cell division capacity. The histogens arise from separate group of initial cells and have different mode of development. These meristematic cells constitute promeristem. These promeritematic cells soon differentiate into three distinct regions (Fig. 2.4) like dermatogen (outer), periblem (middle) and plerome (inner). This each zone consists of a group of distinct initials and each layer called histogens.

1.   Dermatogens: This is an uniseriate outermost layer of the cells which later gives rise to the epidermis of the stem. In root it is also single layered, but at the apex it merge into the periblem and just outside the periblem the dermatogens cuts off many new cells resulting into a small celled tissue called as calyptrogen. The calyptrogens is also meristematic which gives rise to the root cap.

2.   Periblem: It is the middle region composed of isodiametric cells and found in between the dermatogens and the plerome. At the apical region it is single layered but at the central part it is multilayered. The derivatives of these cells produce the cortex of the stem and root.

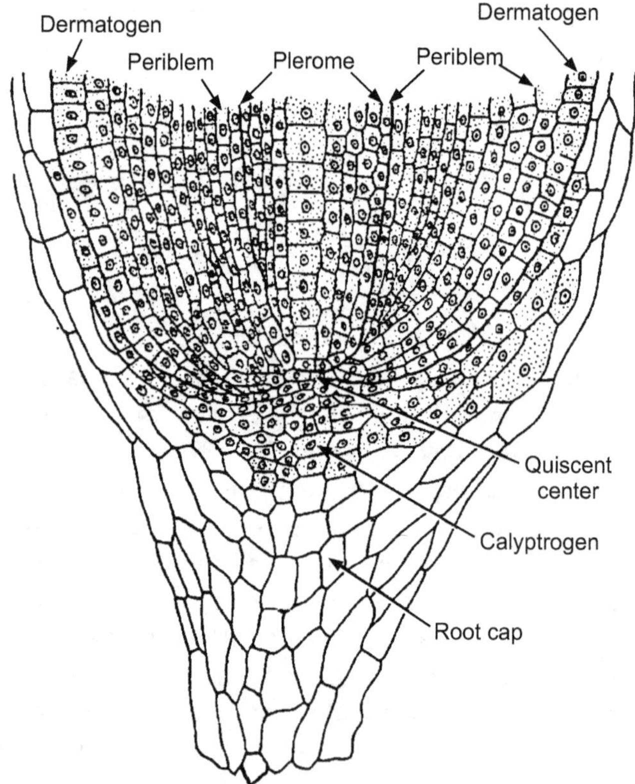

Fig. 2.4: Histogen theory showing different layers

3.   Plerome: It is the massive central meristematic region of the stem apex and lies internal to the periblem. The derivatives of this zone produce the central stele consisting of primary vascular tissues and ground tissues like pericycle, medullary rays and medulla. The cells of this zone mostly multiply in the longitudinal directions.

Further investigations revealed that these layers have no morphological significance and also there is no strict relationship between the development of the histogens and various regions of the plant body. Moreover, it is not possible to distinguish these histogens layers (especially periblem and plerome) in the stem of gymnosperms and angiosperms. Hence this theory was rejected for shoot apical meristem and suggested that it can be applicable only for the root apices in the higher plants.

iii) Tunica Corpus theory: As an interpretation of the apical growth and behavior of meristematic tissues in the shoot apex the third theory i.e. tunica-corpus theory was proposed. Schmidt (1924) has proposed this theory of meristem behavior in the shoot apex of higher plants. The apical cell theory and the histogen theory were developed with reference to shoot and root apex but this theory is proposed only for shoot apex. According to this theory there are two zones of tissue in the apical meristem viz., tunica (cover): consisting of one or more peripheral layers of cell and; Corpus: a mass of cells enclosed by the tunica (Fig. 2.5).

According to this theory there are two distinct zones in the apical meristem of the higher plant body. Tunica is the outer layer consisting of one or more peripheral layers of small uniform cells which normally show anticlinal cell divisions while corpus is the undifferentiated mass of larger cells enclosed by the tunica. The cells of corpus layer vary in number from few to many, divide in many planes and therefore they are more or less irregular in shape and arrangement. Each layer of tunica arises from a group of separate initials and the corpus has one layer of such initials. In the tunica the number of layers of initials is equal to the number of layers of tunica i.e. each layer of tunica has its own layer of initials. The corpus arises from a single tier of initials which divide first periclinally to rise to a group of derivatives, which divide in various plains resulting in the formation of the inner mass of organs. The epidermis arises from the outer layer of tunica while the remaining tissues arise from the corpus (or rarely from tunica).

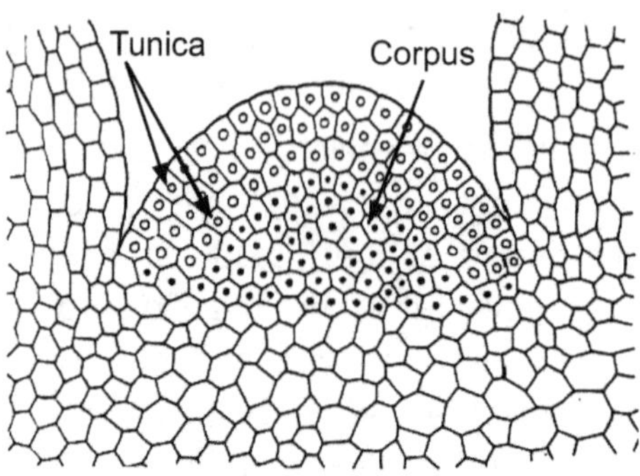

Fig. 2.5: Tunica-Corpus theory showing two zones

Significance of Tunica corpus theory:

1. This theory explains the structural development in stem of primary plant body.

2. It also served well in the establishment of meristematic patterns of the shoot apices of seed plants.

3. By this theory the position, number and behavior of initiating cells in shoot apex of seed plant now much better understood.

4. This theory has topographical value in studies which gives development of lateral organs of the stem like leaves, branches and floral organs.

2.5 PERMANENT TISSUE

The meristematic tissue undergoes cell division and after differentiation produces permanent tissues. The permanent tissues are those that do not have cell division capacity, ultimately these tissues has stopped their growth completely or for the time being. Sometimes these tissues regain the cell division capacity and become meristematic partially or wholly. The tissues may be living or dead and thin-walled or thick walled. Generally the thin walled permanent tissues are living while the thick walled tissue may be living or dead. Permanent tissues may be classified into two main types i.e. Simple and Complex tissues. When the tissues are made up of one types of cells they are called as simple tissues (composed of homogenous cells) while the complex tissues are composed of different types of cells (composed of heterogenous tissues).

2.5.1 Simple Tissue

These tissues are made up of structurally and functionally similar cells and are homogenous in nature. The common simple permanent tissues of the plant are parenchyma, collenchymas and sclerenchyma.

(A) Parenchyma: The parenchyma is the most common tissue present in the plant body. It is composed of living cells which are simple in their morphology and physiology, these tissues are mostly polyhedral in shape (Fig. 2.6). The word parenchyma is derived from the greek 'para' means beside and in-chein means 'to pour' which express the ancient concept of parenchyma as a semi-liquid substance poured beside other tissues which are formed earlier and are more solid. The parenchyma cells are usually isodiametric but they may vary their shape. These cells have intercellular spaces. The cell wall is composed of cellulose or calcium pectate or hemicelluloses.

The Parenchyma tissues make large part of the various organs in many plants like cortex of stem and roots, mesophyll of leaves, the pulp of fruits, endosperm of seeds, pith and in some other organs. The parenchyma tissues also occur in xylem and phloem. The main and primary function of parenchyma is storage of food material.

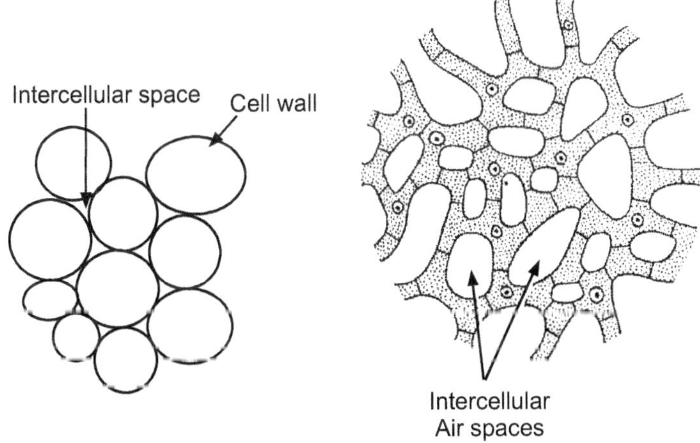

(a) Parenchyma tissues   **(b) Aerenchyma – in the petiole of *Canna***

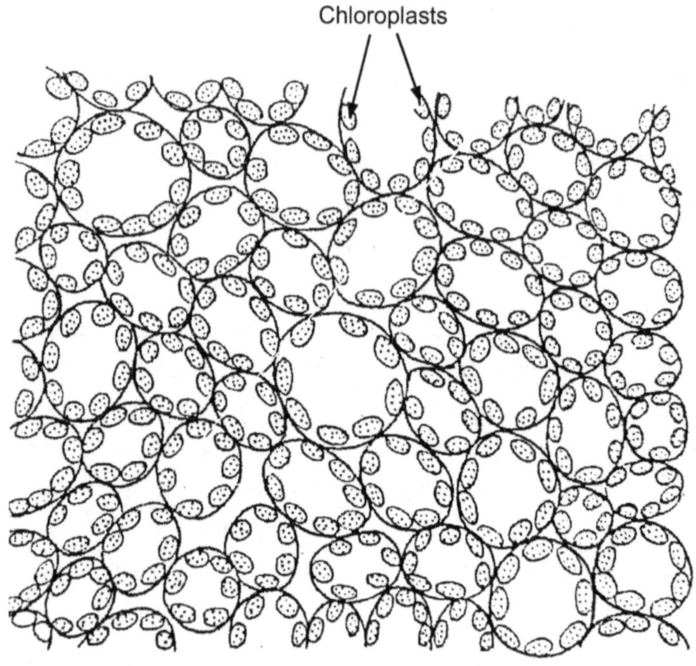

(c) Chlorenchyma

Fig. 2.6

Some modifications are also reported in the parenchyma tissues.

1.   Arenchyma: When the parenchyma tissues possess large air spaces, intercellular spaces, it is referred as arenchyma (Fig. 2.6b). In the aquatic plants the arenchyma tissues are present, these tissues are helpful to float the aquatic plants on the water body. In succulents and xerophytic plants like *Aloe, Agave,* parenchyma may be specialized as water storage tissue.

2.   Chlorenchyma: If the parenchyma cells get exposed to light they develop chloroplasts in them, such tissue is known as chlorenchyma (Fig. 2.6c). It is associated with photosynthesis.

Functions of parenchyma tissues:

In plants parenchyma perhaps the most dominant tissue which carry following functions.

1.   These are associated with storage of food material, it is the primary and most important function.

2. These are associated with various physiological activities like photosynthesis, assimilation, respiration, secretion etc.

3. These are also helpful in the wound healing process of the plants.

4. Parenchyma cells of meristems are helpful in the formation of adventitious buds and roots.

5. Sometimes these tissues regain their cell division capacity and produces secondary tissues.

6. In succulents and some xerophytic plants these tissues works as water storage tissues.

7. In aquatic plants it helpful to float the plant body in the aquatic ecosystem.

8. The photosynthetic parenchymatous tissues are helpful for the photosynthesis.

(B) Collenchyma: Collenchyma is a living tissue composed of elongated cells with thick primary walls. Morphologically it is simple tissue and consists of one type of cell. Important characteristic feature of this tissue is that these tissues do not possess intercellular spaces (compactly arranged cells) and are thickened at the corners of the cells (Fig. 2.7). The thickening is due to deposition of cellulose or hemicelluloses or protopectin but not lignin.

These are present in aerial parts of dicot plants and absent in roots and monocots. Collenchyma tissues are mostly present in the peripheral regions of rapidly elongating organs like young stem, leaf petiole, stalk of the flower and the leaves. As these cells are very flexible in nature they have adaptability to change their shaped in the rapidly growing organ, especially those of increase in length. It is the only living mechanical tissue which gives physical strength or mechanical support mostly to herbaceous plants. As it possess chloroplast it also manufactures sugar and starch, so the collenchyma shows two functions i.e. it gives mechanical strength to the plant body and photosynthesis.

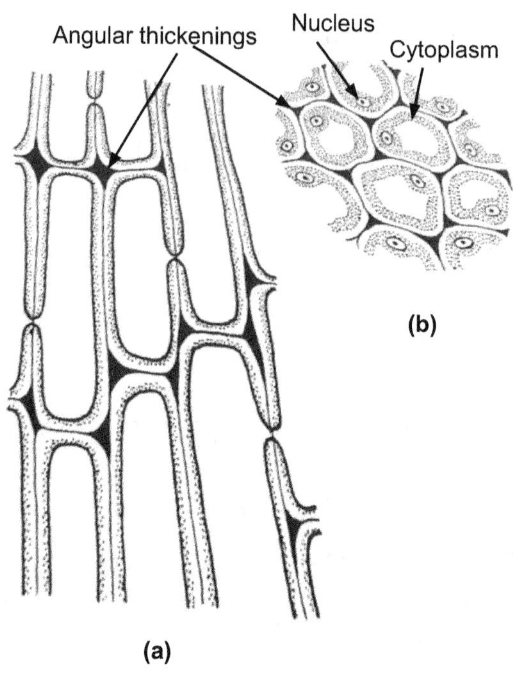

Fig. 2.7: Structure of Collenchyma

(C) Sclerenchyma: The word Sclerenchyma is derived from two Latin words, 'Scleros- means hard and 'enchyma means an infusion'. It is usually with pointed ends and consists of very long, narrow, thick and lignified cells. It can be broadly grouped into two types i.e. Sclerenchyma fibres and Sclereids.

(i) Sclerenchyma fibres: As they are fibre-like in appearance, are called as sclerenchyma fibres or simply fibres. They have simple, often oblique, pits in their walls (Fig. 2.8). Sclerenchymatous cells are abundantly found in plants and distributed in the patches or definite layers. Rarely, they also occur singly associated with other cells. These are permanent tissues and dead in nature. These are purely mechanical tissues which gives mechanical strength to the different plant organs. It gives the requisite strength, rigidity, flexibility and elasticity to the plant body. They may be variable in length, in angiosperms the average is about 1 to 3 mm, 2 to 8 mm in gymnosperms but in some special cases like in hemp (*Cannabis*), rhea (*Boehmeria nivea*), flax (*Linum*) etc. the fibres are of excessive lengths, ranging from 20 to 550 mm in length. These thick walled and rigid cells constitute the excellent textile fibres that have commercial use. The fibres of jute, coconut, Indian or sunn hemp (*Crotalaria juncea*), Madras or deccan hemp (*Hibiscus cannabis*), sisal hemp (*Agave sisalana*) and many others are the common examples of long fibres.

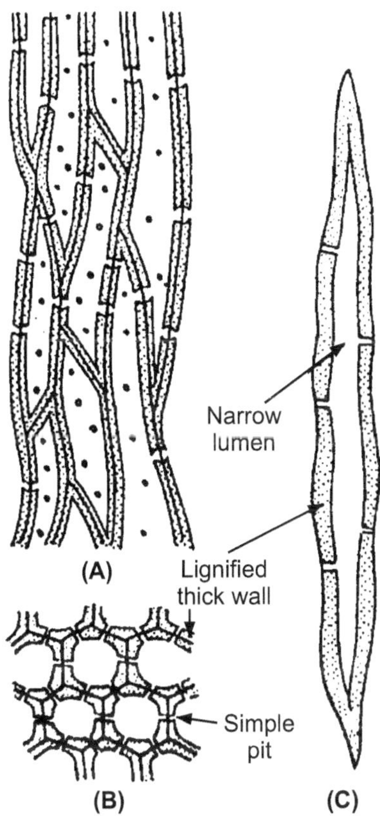

Fig. 2.8: Structure of Sclerenchyma Fibre

(ii) Sclereids: These are widely distributed in the plant body. These are special types of sclerenchyma developed in various parts of the plant body to meet local mechanical needs. They are also known as sclerotic cells. They are usually much longer than the broad, occurring singly or in groups. Mostly these are isodiametric in shape but sometimes they may be elongated. They may occur in the cortex, pith, phloem, hard seed, nuts, stony fruits and in the leaves and stems of many dicotyledons and also gymnosperms. They are most common in fruits and seeds. In fruits they are distributed singly or in groups which gives hardness and strength to the fruit wall or seed coat.

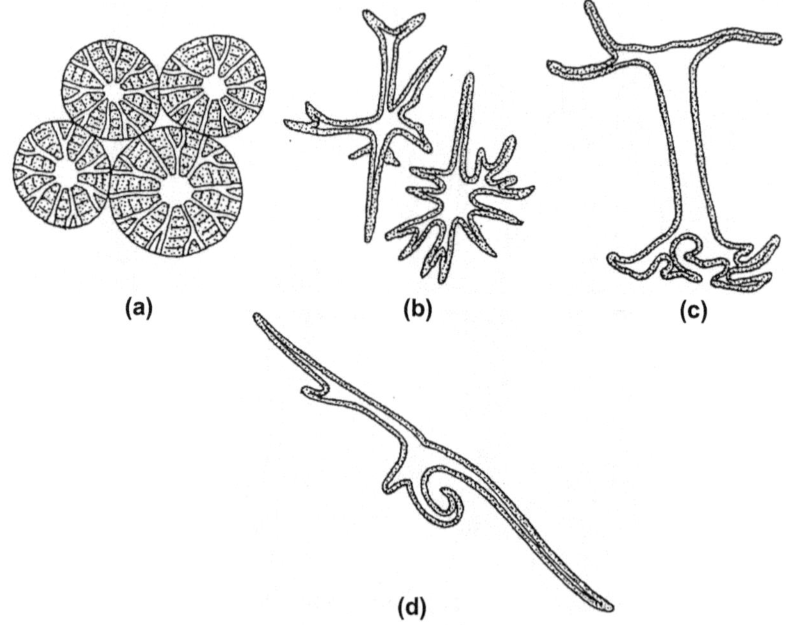

Fig. 2.9: Different types of sclereids: (a) Brachysclereids, (b) astrosclereids, (c) Osterosclereid and (d) Trichosclereid

Sclereids are further divided into various types like:

1. Brachysclereids (Brachys = short): These are also known as stone cells. These are more or less isodiametric in nature (Fig. 2.9a) and are commonly found among masses of parenchyma in different parts of the plant body like in the pith, cortex, bark, phloem, fleshy pericarp of certain fruits, coconut shell and many others.

2. Macrosclereids (makris = long): These are rod-like or columnar cells with truncated ends, forming a solid, palisade-like epidermal layer. These tissues are mostly present in the members of family Leguminosaceae and mostly distributed in the bark and seed-coats. The protective scales of garlic and onion also have these tissues (Fig. 2.9d).

3. Osteoclereids (ostoon = bone): They are also called as columnar cells. They are dilated or lobed at both ends and looks like somewhat bones (Fig. 2.9c). They are commonly found in the seed-coats and fruit walls and also in the leaves of some dicotyledonous plants.

4. Astroclereids (astron = star): These are irregular stone cells giving stellate or star-like appearance (Fig. 2.9b). Generally they are found in the leaves of certain dicotyledonous (leaves of tea) and also in few gymnosperms like *Tsuga* (a conifer) and *Gnetum*. They are also reported from the bark of *Abies* and *Larix* etc.

2.5.2 Complex Tissue

These are heterogenous in nature as they are composed of different types of cell elements. Xylem and phloem are the complex tissues which constitute the component parts of the vascular bundle and hence called as vascular tissues. The vascular bundles form a continuous and interconnected system in the different plant organs of the plant body. The vascular system occupies a unique position in the plant body. It is primarily associated with transport of water, solutes and food material.

A. Xylem: It is a complex tissue mainly associated with conduction of water and solutes. As it is complex tissues, it is composed of different types of cells and elements including living and non-living. These tissues are tracheids, trachea or vessels, xylem fibres or wood fibres and xylem parenchyma (wood parenchyma). Out of these tissues xylem parenchyma is the only living in nature and remaining are dead. The main components of xylem are given in Fig. 2.9:

1. Tracheids: A tracheid is a fundamental cell type in xylem. These are very much elongated cells occurring along the long axis of organ. The cells of tracheids are without protoplast and they are dead by nature. It is an elongated tube-like cell having tapering, rounded or oval ends and hard and lignified walls. In cross section tracheids are round or polyhedral. Tracheid is the only type of element found in the fossils of seed-plants. Tracheids are dominant in pteridophytes and gymnosperms but also present in wood of angiosperms.

The wall is hard, moderately thick and usually lignified. The wall of tracheids posses various type of thickenings in them and they may be annular, spiral, reticulate, scalariform or pitted. But pits of the bordered types are most abundant. These pits are important to establish communication with adjoining tracheids and also with other cells. Due to the presence of central lumen and hard lignified wall tracheids are nicely adapted for transport of water and solutes. They also serve as supporting tissue.

2. Vessels: It is also known as Tracheae. These are long tube-like structures ideally suited for the conduction of water and solutes. A vessel is formed from a row of cylindrical cells arranged in longitudinal series where the partition walls become perforated, so that the whole thing serves like a tube. As opening of the vessels are distinct and therefore translocations of solutes become easier in the vessels. The vessels are considerably long bodies, it varies from few centimeter to few meters (in ash plants, *Fraxinus excelsior* of family Oleaceae vessels are ca. ten feet). Like tracheids, vessels are also devoid of protoplast and hence it is dead in nature and also have different types of thickenings.

Tracheids	Vessels
1. Tracheids are short and generally upto 1 mm in length.	1. Vessels are comparatively longer and generally ca. 10 cm in length.
2. It is composed of a single elongated cell with tapering ends.	2. It is composed of a row of cells placed one above the other.
3. The tracheids are not tubular.	3. Vessels are tubular in shape.
4. Tracheids found one above the other are separated by cross walls.	4. Vessels do not have cross walls.

3. Xylem fibres: Some fibres remain associated with other elements in the complex tissue and mainly give mechanical support to the plant body. The fibres are mainly of two types i.e. tracheids and libriform fibres. It is very difficult to separate these two types. But fibre-tracheids posses bordered pits, though the borders are not well-developed while libriform fibres are narrow ones with highly thickened walls. The central lumen is almost obliterated and pits are simple. They occur abundantly in many woody dicot plants.

4. Xylem parenchyma: It is the only living tissue among the xylem elements. The cells may be thin-walled or thick-walled. In primary xylem it is remain associated with other elements. In secondary xylem parenchyma occurs in two forms: Xylem parenchyma which is somewhat elongated cell and mostly lies in vertical series are attached end to end while ray parenchyma cells occur in radial transverse series in many woody plants. The xylem parenchyma is abundant in the secondary xylem of most of the plants excepting few conifers like *Pinus*, *Taxus* and *Araucauria*.

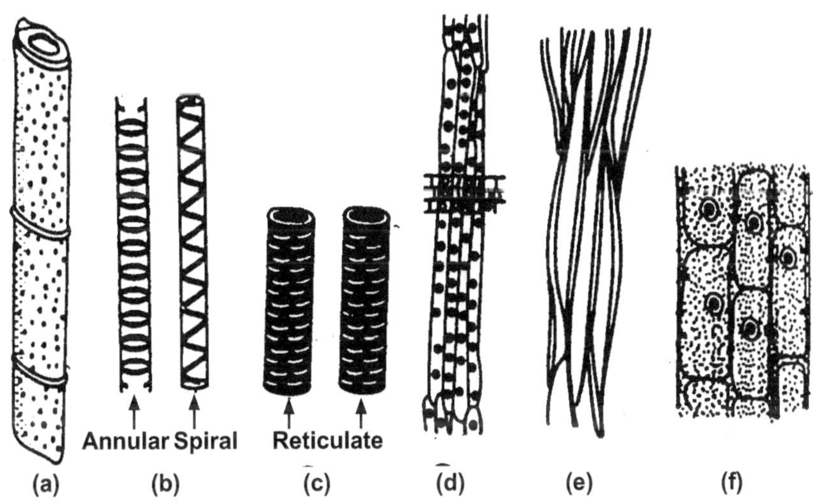

Fig. 2.9: Elements of Xylem:(a) Vessel tube, (b) & (c) Thickening on vessels, (d) Tracheids, (e) Xylem fibre, (f) Xylem parenchyma

These cells are mostly associated with storage of starch and fatty food, sometimes also store other matters like tannins, crystals etc. As it is a part of xylem they are possibly involved in conduction of water and solutes and also gives mechanical support to the plant.

B. Phloem: The phloem or bast is another conducting tissues and is composed of the following elements; a. Sieve-tubes, b. Companion cells, c. Phloem parenchyma, d. Bast fibres (rarely).

(Fig. 2.10) Phloem, as a whole, is meant to conduct prepared food materials from leaf to the storage organs and growing regions.

1. Sieve tubes: Sieve tubes are slender, tube-like structures, composed of elongated cells which are places end to end. Their walls are thin and made up of cellulose. The transverse partition walls are however perforated by a number of pores. The transverse wall then looks very much like a sieve and then is called the sieve-plate. The sieve-plate may sometimes be formed in the side (longitudinal) wall. In some cases, the sieve plate is not transverse (horizontal) but incline obliquely and then different areas of it become perforated. A sieve plate of this nature is called a compound plate. At the close of the growing season, the sieve plate is covered by a deposit of colourless, shining substance in the form of a pad, called the callus or callus pad. This consists of carbohydrate called callose. In winter, the callus completely clogs the pores, but in spring, when the active season begins, it gets dissolved. In old sieve tubes the callus forms a permanent deposit. The sieve tube contains no nucleus, but has a lining layer of cytoplasm, which is continuous through the pores. Sieve tubes are used for the longitudinal transmission of prepared food materials- proteins and carbohydrates- downward from the leaves to the storage organs and later upward from the storage organs the growing regions. A heavy deposit of food material is found on either side of the sieve tube with a narrow median portion.

2. Companion cells: It is associated with each sieve tube and connected with pores known as the companion cell. It is living and contains protoplasm and an elongated nucleus. The companion cell is present only in angiosperms (both dicots and monocots). It assist the sieve tube in the conduction of food.

3. Phloem parenchyma: There are always some parenchymatous cells from a part of the phloem in all dicotyledons, gymnosperms and ferns. These cells are living, they often cylindrical. They store up food material and help to conduct it. Phloem parenchyma is however, absent in most monocotyledons.

4. Bast fibres: Sclerenchymatous cells occurring in the phloem and bast are known as the bast fibres. These are generally absent in the primary phloem, but occur frequently in the secondary phloem.

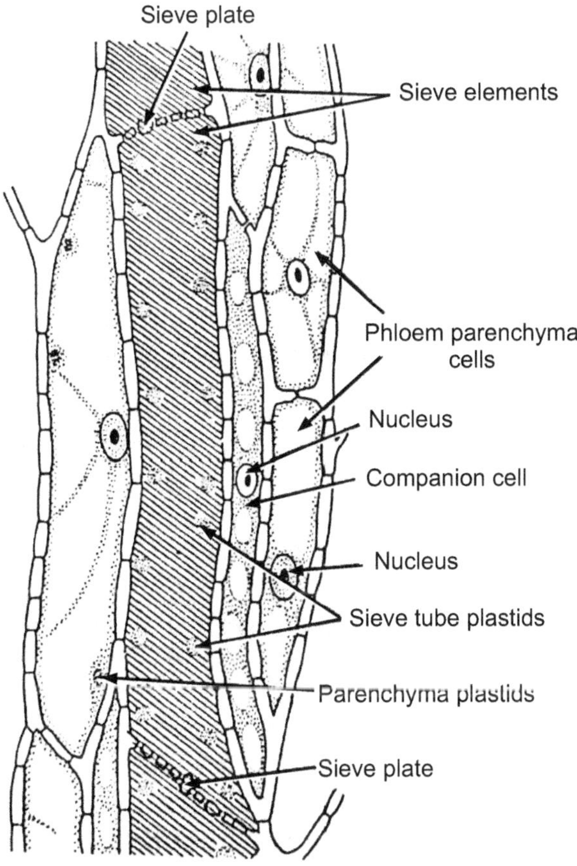

Fig. 2.10: Elements of Phloem

2.6 TYPES OF VASCULAR BUNDLES

Plants perform the function of photosynthesis, for this process it requires water, minerals, solutes, these substances are obtained from root; after photosynthesis starch (carbohydrate) is produces which is stored into different plant organs. In plants the translocation of these substances is performed by some specific tissues and these tissues are known as conducting tissues and also vascular tissues. The system is composed of such tissues is known as vascular tissue system or conductive tissue system. The vascular tissue system consists of complex permanent tissues like xylem and phloem and these tissues are bounded in a bundle and hence called as vascular bundle. The vascular tissues usually perform the functions like conduction of water and mineral salts and organic food material from one region to another in different directions. The xylem is associated with conduction of water and dissolved minerals while phloem with conduction of food material. Usually xylem and phloem elements are placed in fascicles (fascis means bundle) and hence called vascular bundle. In such vascular elements the xylem and phloem may be separated with a cambium strip (cells of this strip are meristematic in nature) or may not be separated by such strip. On the nature of arrangement of vascular tissues, the vascular bundles show following three common types of arrangement: (i) The two complex tissues lies side by side, (ii) One tissue remain surrounded by other tissue, (iii) The two tissues are separated from each other. On the basis of arrangement of xylem and phloem in the vascular bundle they are divided into different types.

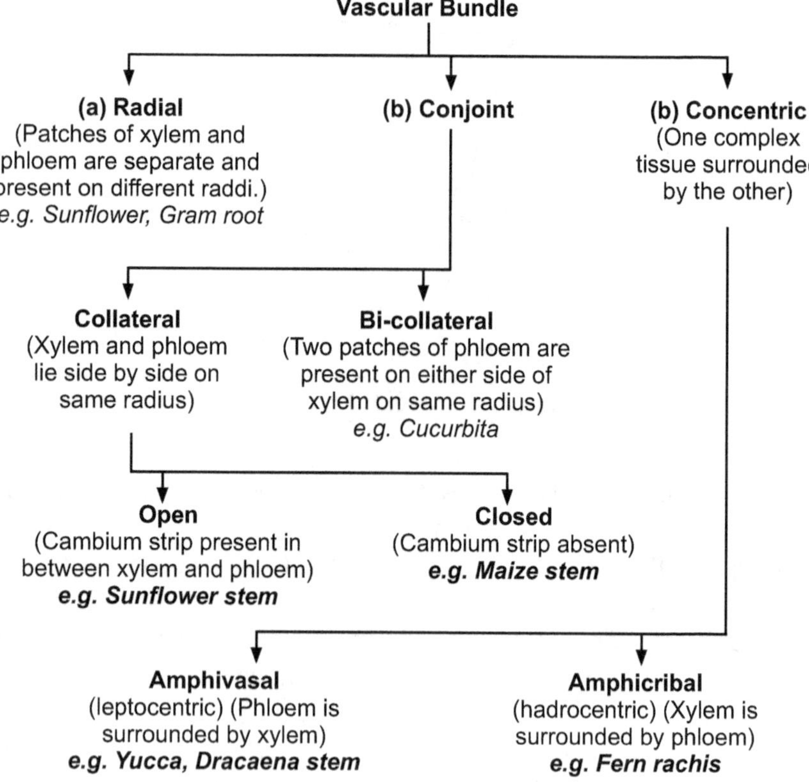

a) Radial vascular bundle: In this type xylem and phloem present on the alternate radii. They are often separated by patches of non-conductive tissues like parenchyma and sclerenchyma tissues. In all roots the vascular bundles are of radial type (Fig. 2.11). These types of vascular bundles are always closed type bundles, as there is no cambium between xylem and phloem. These are further classified into various types on the basis of position of protoxylem and metaxylem and also on the basis of number of xylem patches. In this type xylem is always exarch i.e. protoxylem is towards periphery.

b) Conjoint vascular bundle: In this type xylem and phloem are present on the same radius and lies side by hence called as conjoint type vascular bundle. On the basis of number of phloem patches in the bundle it is further divided into two types like -

1. Conjoint collateral vascular bundle: In this type, the xylem and phloem tissues remain side by side and arranged on the same radius. In stem such bundles shows that xylem is endarch (i.e. protoxylem towards the centre and metaxylem towards peripheri) (Fig. 2.11). It is open type because of presence of cambium (lateral meristem) between xylem and phloem. In most of the stems of dicot plants the vascular bundles are conjoint, collateral and open while in stem of monocot plants it is conjoint, collateral and closed type because there is absence of cambium strip in between xylem and phloem.

2. Conjoint bi-collateral vascular bundle: Such vascular bundles are present in majority members of family Cucurbitaceae and few members of Solanaceae, Convolvulaceae, Myrtaceae, Apocynaceae etc. Such type of vascular bundles are always open due to the presence of cambium strips. In this type of vascular bundles two phloem patches are present on either side of the xylem patch. The phloem outside the xylem is called outer phloem and that on its inner side is called inner phloem (Fig. 2.11). On both the sides, between xylem and phloem, cambium strips are present these cambial strips are called outer cambium and inner cambium respectively. Thus the vascular bundle possesses two phloem patches and two cambium strips but with the single xylem patch therefore such vascular bundle is known as bi-collateral.

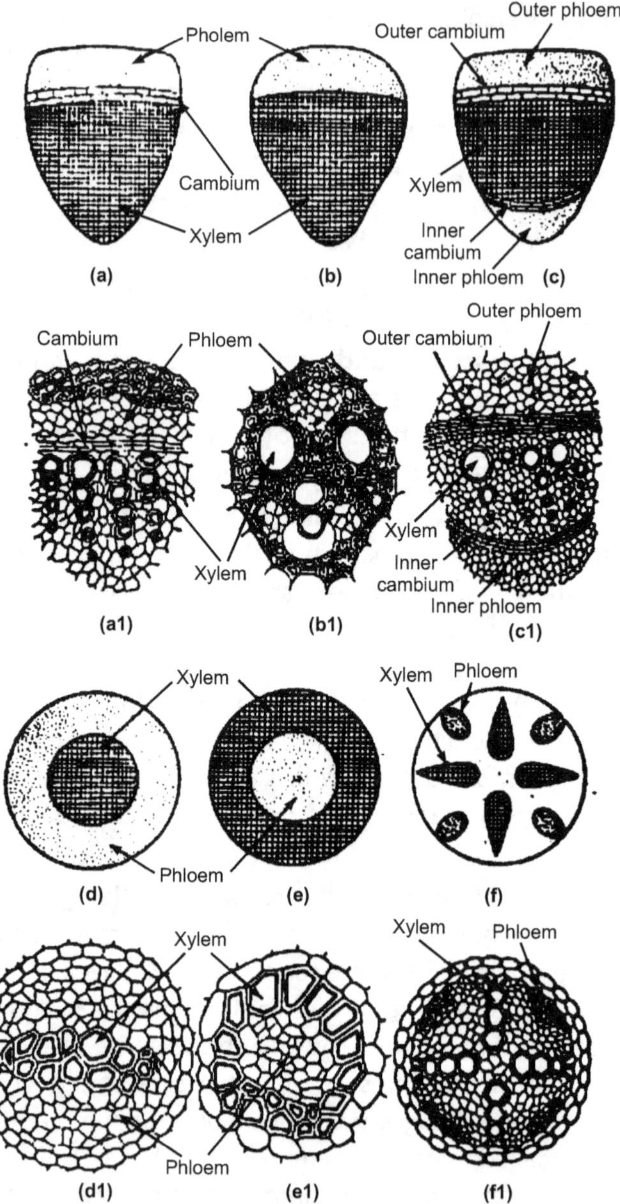

Fig. 2.11: Types of Vascular bundles: (a) Model of collateral open bundle, (a1) Same in t.s. (b) Collateral closed (model), (b1) Same in t.s. (c) Bicollateral (model), (c1) Same in t.s. (d) Concentric-amphicribral (model), (d1) Same in t.s. (e) Concentric-amphivasal (moel). (e1) Same in t.s. (f) Radial (model), (f1) Same in t.s.

c) Concentric vascular bundle: When one type of vascular tissue (may be xylem or phloem) surrounds the other, the bundles are known as concentric vascular bundles. These types further divided into two sub-types viz.,

1. Amphivasal (Leptocentric) bundle: In this type of vascular bundle phloem is surrounded by xylem, it is also referred as leptocentric (leptos-phloem, centric-in centre) (Fig. 2.11). The common examples of this type are stem of *Dracaena, Yucca* etc.

2. Amphicribal (Hydrocentric) bundle: In this type the xylem is surrounded by the phloem. It is also known as hydrocentric (hydro = xylem, centric = in the centre). These types of vascular bundles are present in ferns rachis, pedicel of flowers and fruits (Fig. 2.11). These types of vascular bundles are always closed type due to absence of cambium in between xylem and phloem.

QUESTIONS

A. Multiple Choice Questions:

1. Cells that are having cell division capacity are known as........................ cells.

    (a) permanent   (b) meristematic

    (c) epidermal   (d) secretary

2. The lateral meristem is responsible for to increase ....

    (a) height   (b) thickness

    (c) size of parenchyma (d) length of leaves

3. The intercalary meristems are mostly found in ....

    (a) parenchyma   (b) collenchyma

    (c) cork cambium   (d) chlorenchyma

4. Cork cambium is .... merlstem.

    (a) lateral   (b) apical

    (c) intercalary   (d) pro

5. Vascular cambium is an example of .... meristem.
   - (a) primary
   - (b) apical
   - (c) tertiary
   - (d) secondary

6. The living mechanical tissue is ....
   - (a) chlorenchyma
   - (b) collenchyma
   - (c) arenchyma
   - (d) sclerids

7. Bicollateral vascular bundles are mostly found in members of family ....
   - (a) Cucurbitaceae
   - (b) Poaceae
   - (c) Magnoliaceae
   - (d) Liliaceae

8. Open type of vascular bundles are generally found in ....
   - (a) monocot stem
   - (b) dicot stem
   - (c) dicot root
   - (d) fern rachis

9. Closed type of vascular bundles are generally found in ....
   - (a) monocot stem
   - (b) dicot stem
   - (c) dicot root
   - (d) fern rachis

10. When vascular bundles have cambium in between xylem and phloem, it is referred as.........
    - (a) open
    - (b) closed
    - (c) concentric
    - (d) leptocentric

11. .... tissues are dead in nature.
    - (a) meristematic
    - (b) permanent
    - (c) xylem parenchyma
    - (d) root cap

12. Cork cambium is an example of .... meristem.
    - (a) primary
    - (b) apical
    - (c) tertiary
    - (d) secondary

13. Apical cell theory was first time proposed by ....
    - (a) Hanstein
    - (b) Nageli
    - (c) Hofmeister
    - (d) Strasburger

14. .... theory is only applicable for lower plants.
    (a) Histogen
    (b) Tunica Corpus
    (c) Apical cell
    (d) Wolff

15. Histogen cell theory was first time proposed by ....
    (a) Hanstein
    (b) Nageli
    (c) Hofmeister
    (d) Strasburger

16. In Histogen cell theory, derivatives of periblem produces........................
    (a) epidermis
    (b) cortex
    (c) xylem
    (d) phloem

17. In Histogen cell theory, derivatives of plerome produces........................
    (a) only xylem
    (b) cortex
    (c) xylem and phloem
    (d) only phloem

18. In root apex .... is responsible for formation of root cap.
    (a) xylem
    (b) cortex
    (c) calyptrogen
    (d) dermatogen

19. Tunica Corpus theory was first time proposed by ....
    (a) Hanstein
    (b) Schmit
    (c) Hofmeister
    (d) Strasburger

20. In Tunica Corpus theory, there are ...... layers in the growing apices of higher plant body.
    (a) one
    (b) two
    (c) three
    (d) four

21. In Tunica Corpus theory, ... is responsible for formation of main bulk of plant body.
    (a) corpus
    (b) tunica
    (c) periblem
    (d) plerome

22. ...... tissue is the most common tissue in the plant body.
    (a) parenchyma
    (b) sclerenchyma
    (c) collenchyma
    (d) xylem

23. When parenchyma tissues have air chambers, it is referred as
....

   (a) parenchyma   (b) sclerenchyma

   (c) collenchyma   (d) arenchyma

24. When parenchyma tissues possess chloroplast, it is referred as
....

   (a) parenchyma   (b) chlorenchyma

   (c) collenchyma   (d) arenchyma

25. The primary function of parenchyma is ....

   (a) storage of food   (b) photosynthesis

   (c) mechanical support (d) anchorage

26. In herbaceous plants .... tissue gives mechanical support
to the plant body.

   (a) parenchyma   (b) chlorenchyma

   (c) collenchyma   (d) arenchyma

27. .... tissues do not possess intercellular spaces.

   (a) parenchyma   (b) chlorenchyma

   (c) collenchyma   (d) arenchyma

28. Fibre of jute is the example of .... tissues.

   (a) sclerenchyma fibre

   (b) chlorenchyma

   (c) collenchyma

   (d) arenchyma

29. Among the xylem elements .... tissues is only living
tissue.

   (a) xylem parenchyma

   (b) xylem fibre

   (c) xylem vessel

   (d) xylem tracheid

30. .... tissues are only present in the angiosperms.

    (a) Xylem parenchyma

    (b) Companion cells

    (c) Phloem parenchyma

    (d) Fibres tracheid

Answer key:

(1) – b, (2) – b, (3) – c, (4) – c, (5) – a, (6) – b, (7) – a, (8) – b, (9) – a, (10) – a, (11) – b, (12) – d, (13) – a, (14) – c, (15) – a, (16) – b, (17) – c, (18) – c, (19) – b, (20) – b, (21) – a, (22) – a, (23) – d, (24) – b, (25) – a, (26) – c, (27) – c, (28) – a, (29) – a, (30) – b.

B. Broad Questions:

1. Explain meristem. Describe types of meristem based on position.

2. Describe in brief the structural development of plants with histogen theory.

3. Explain with suitable diagrams Tunica Corpus theory for structural development of higher plant body.

4. Describe with diagrams types of vascular bundles.

5. What is tissue? Explain different types of simple tissues in higher plants.

6. What is tissue? Explain different types of complex tissues in higher plants.

7. What is tissue? Explain different types of Sclerenchyma tissues in

   higher plants.

8. Explain in details structure of xylem and phloem.

9. Explain structure and functions of meristem and parenchyma tissues in higher plants.

10. Discuss in brief apical and histogen theory of structural development in plants.

C. **Write Short Notes on:**

1. Meristem- structure and functions
2. Apical cell theory
3. Parenchyma tissues
4. Functions of parenchyma tissues
5. Collenchyma
6. Sclerenchyma fibres
7. Types of sclereids
8. Xylem
9. Phloem
10. Difference between tracheids and vessels
11. Sieve tubes
12. Conjoint vascular bundles
13. Bicollateral vascular bundle
14. Types of concentric vascular bundles
15. Xylem parenchyma

PRIMARY AND SECONDARY STRUCTURE OF PLANT BODY

3.1 PRIMARY STRUCTURE OF MONOCOTYLEDON AND DICOTYLEDON ROOT, STEM AND LEAF

The embryo develops into an adult plant with roots, stem and leaves due to the activity of the apical meristem. A mature plant has three kinds of tissue systems - the dermal, the fundamental and the vascular system. The dermal system includes the epidermis, which is the primary outer protective covering of the plant body. The periderm is another protective tissue that supplants the epidermis in the roots and stems that undergo secondary growth. The fundamental tissue system includes tissues that form the ground substance of the plant in which other permanent tissues are found embedded. Parenchyma, collenchyma and sclerenchyma are the main ground tissues. The vascular system contains the two conducting tissues, the phloem and xylem. In different parts of the plants, the various tissues are distributed in characteristic patterns. This is best understood by studying their internal structure by cutting sections (transverse or longitudinal or both) of the part to be studied.

3.1.1 Primary Structure of Monocotyledon Root, Stem and Leaf

(1) Primary Structure of Monocotyledonous Root - Maize Root

The internal structure of the monocot roots shows the following tissue systems from the periphery to the centre. They are epidermis, cortex and stele.

Epidermis: It is the outermost layer of the root. It is also known as rhizodermis or epiblema. It consists of a single row of thin-walled parenchymatous cells without any intercellular space. Stomata and cuticle are absent in the rhizodermis. Root hairs that are found in the rhizodermis are always unicellular. They absorb water and mineral salts from the soil. Root hairs are generally short lived. The main function of rhizodermis is protection of the inner tissues.

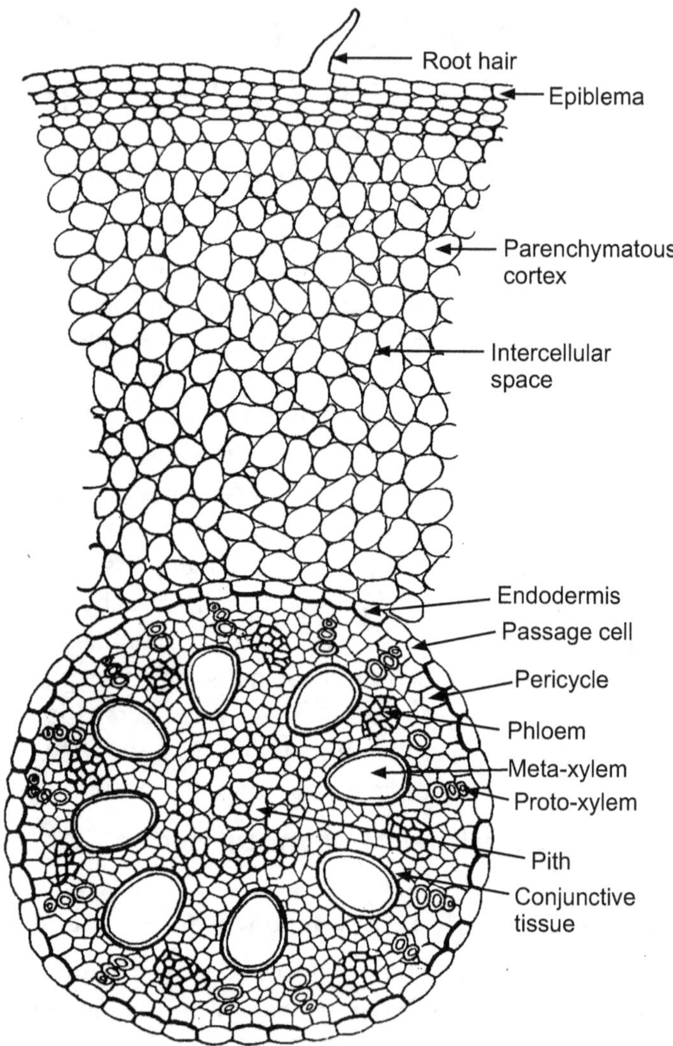

Fig. 3.1: T. S. of Maize root showing primary structure

Cortex: The cortex is homogenous. i.e. the cortex is made up of only one type of tissue called parenchyma. It consists of many layers of thin-walled parenchyma cells with lot of intercellular spaces. The function of cortical cells is storage. Cortical cells are generally oval or rounded in shape. Chloroplasts are absent in the cortical cells, but they store starch. The cells are living and possess leucoplasts. The innermost layer of the cortexis endodermis. It is composed of single layer of barrel shaped parenchymatous cells. This forms a complete ring around the stele. There is a band like structure made of suberin present in the radial and transverse walls of the endodermal cells. They are called Casparian strips named after Casparay who first noted the strips. The endodermal cells, which are opposite to the protoxylem elements are thin-walled without casparian strips. These cells are called passage cells. Their function is to transport water and dissolved salts from the cortex to the xylem. Water cannot pass through other endodermal cells due to casparian strips. The main function of casparian strips in the endodermal cells is to prevent the re-entry of water into the cortex once water entered the xylem tissue.

Stele: All the tissues inside the endodermis comprise the stele. This includes pericycle, vascular system and pith.

Pericycle: Pericycle is the outermost layer of the stele and lies inner to the endodermis. It consists of a single layer of parenchymatous cells.

Vascular System: Vascular tissues are seen in radial arrangement. The number of protoxylem groups is many. This arrangement of xylem is called polyarch. Xylem is in exarch condition. The tissue, which is present between the xylem and the phloem, is called conjunctive tissue. In maize, the conjunctive tissue is made up of sclerenchymatous tissue.

Pith: The central portion is occupied by a large pith. It consists of thin walled parenchyma cells with intercellular spaces. These cells are filled with abundant starch grains.

(2) Primary Structure of Monocotyledon Stem - Maize Stem

The outline of the maize stem in transverse section is more or less circular. Internal structure of monocotyledonous stem reveals epidermis, hypodermis, ground tissue and vascular bundles.

Epidermis: It is the outermost layer of the stem. It is made up of single layer of tightly packed parenchymatous cells. Their outer walls are covered with thick cuticle. The continuity of this layer may be broken here and thereby the presence of a few stomata. There are no epidermal outgrowths.

Hypodermis: A few layer of sclerenchymatous cells lying below the epidermis constitute the hypodermis. This layer gives mechanical strength to the plant. It is interrupted here and there by chlorenchyma cells.

Ground tissue: There is no distinction into cortex, endodermis, pericycle and pith. The entire mass of parenchymatous cells lying inner to the hypodermis forms the ground tissue. The cell wall is made up of cellulose. The cells contain reserve food material like starch. The cells of the ground tissue next to the hypodermis are smaller in size, polygonal in shape and compactly arranged. Towards the centre, the cells are loosely arranged, rounded in shape and bigger in size. The vascular bundles lie embedded in this tissue. The ground tissue stores food and performs gaseous exchange.

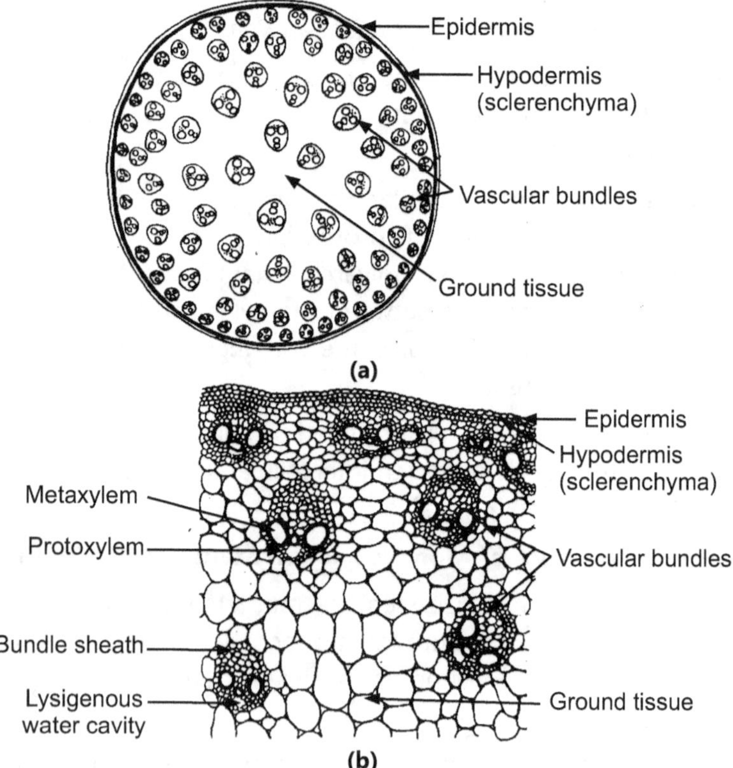

Fig. 3.2: Primary structure of Monocotyledon stem - Maize stem

Vascular bundles: Vascular bundles are scattered in the parenchymatous ground tissue. Each vascular bundle is surrounded by a sheath of sclerenchymatous fibres called bundle sheath. The vascular bundles are conjoint, collateral, endarch and closed. Vascular bundles are numerous, small and closely arranged in the peripheral portion. Towards the centre, the bundles are comparatively large in size and loosely arranged. Vascular bundles are skull shaped.

Phloem: The phloem in the monocot stem consists of sieve tubes and companion cells. Phloem parenchyma and phloem fibres are absent.

Xylem: Xylem vessels are arranged in the form of the letter 'Y'. The two metaxylem vessels are located at the upper two arms and one or two protoxylem vessels at the base. In a mature bundle, the lowest protoxylem disintegrates and forms a cavity known as lysigenous water cavity or protoxylem lacuna.

(3) Primary Structure of a Monocotyledon leaf –Maize leaf

A transverse section of a maize leaf shows the following internal structures;

Epidermis: Both upper and lower epidermal layers are uniseriate and composed of more or less oval cells with a cuticle layer on the leaf surface. The upper epidermis contains large and empty bulliform cells. Stomata are present on both the epidermal layers.

Mesophyll: The mesophyll is not differentiated into palisade and spongy parenchyma. All the mesophyll cells are compactly arranged isodiametric and thin walled. These cells are compactly arranged with limited intercellular spaces. They contain numerous chloroplasts.

Vascular bundles: Vascular bundles are conjoint, collateral and closed. Vascular bundles differ in size. Most of the vascular bundles are smaller in size. Large bundles occur at regular intervals. Two patches of sclerenchyma are present above and below the large vascular bundles. These sclerenchyma patches give mechanical support to the leaf. The small vascular bundles do not have such sclerenchymatous patches. Each vascular bundle is surrounded by a parenchymatous bundle sheath. The cells of the bundle sheath generally contain starch grains. Maize is a C4 plant where the Hatch and Slack pathway operates to concentrate CO_2 in the bundle sheath cells to minimize photorespiration. Inside the bundle sheath cells Calvin cycle operates. The xylem of the vascular bundle is located towards the upper epidermis and the phloem towards the lower epidermis.

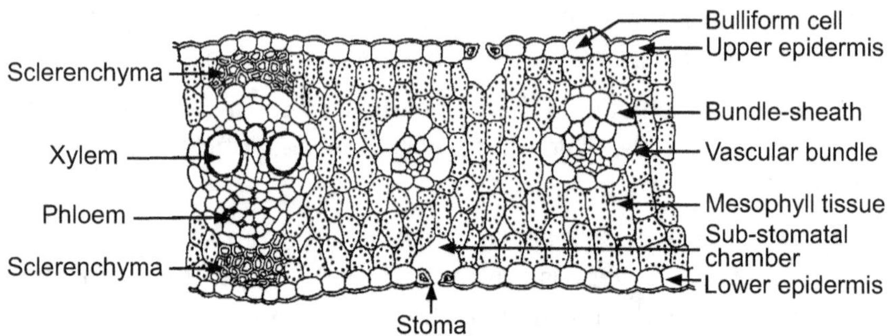

Fig. 3.3: Primary Structure of a Monocotyledon leaf – Maize leaf

3.1.2 Primary Structure of Dicotyledon Root, Stem and Leafs

(1) Primary structure of dicotyledonous root - Sunflower root (*Helianthus annuus*)

The transverse section of the dicot root (Sunflower root) shows the following arrangement of tissues:

Epidermis: The outermost layer of the root is also known as *rhizodermis* or *epiblema*. It is made up of a single layer of parenchyma cells which are arranged compactly without intercellular spaces. It is devoid of stomata and cuticle. Root hair is always single celled. It absorbs water and mineral salts from the soil. The chief function of rhizodermis is protection.

Cortex: Next to epidermis there is a homogenous massive parenchymatous zone called the cortex consisting of parenchyma cells with conspicuous intercellular spaces. The cell contain large amount of leucoplasts. Their main function is to translocate water and minerals absorbed by the root hairs to the conducting elements. The innermost layer of cortex is the endodermis consisting of compactly arranged barrel shaped cells forming a distinct cell layer surrounding the stele. The cells of this layer possess Casparian strips on their radial walls.

Stele: All the tissues present inside endodermis comprise the stele. It includes pericycle and vascular system.

Pericycle: It is generally a single layer of parenchymatous cells found inner to the endodermis. It is the outermost 'layer of the stele. Lateral roots originate from the pericycle. Thus, the lateral roots are endogenous in origin.

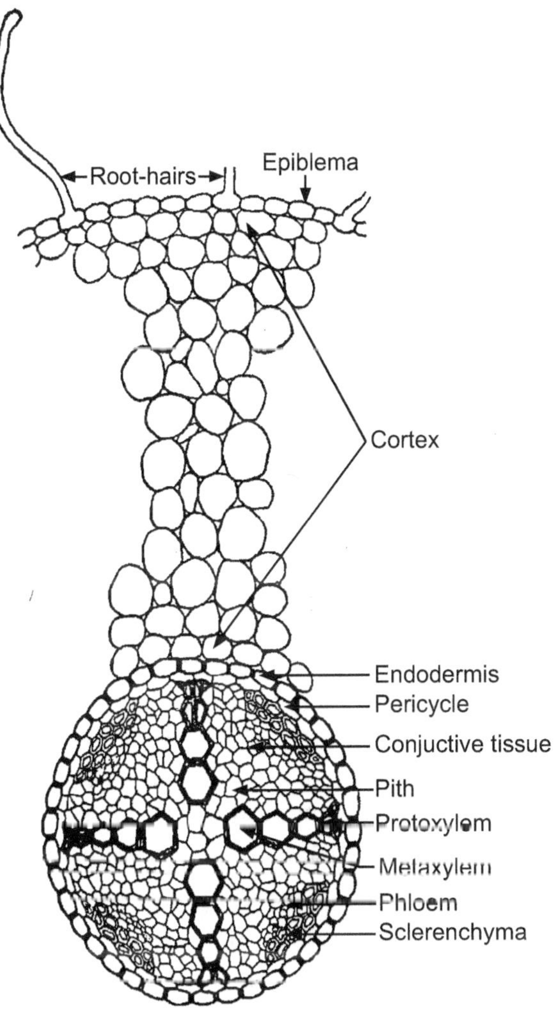

Fig. 3.4: Primary structure of dicotyledonous root - Sunflower root

Vascular system: Vascular tissues are in radial arrangement. The tissue by which xylem and phloem are separated is called conjunctive tissue. The conjunctive tissue is composed of parenchymatous tissue. Xylem is inexarch condition. The number of protoxylem points is four and so the xylem is called *tetrarch*. Each phloem patch consists of sieve tubes, company. In the centre parenchymatous pith is present.

(2) Primary structure of dicotyledonous stem - Sunflower stem

Internal structure of dicotyledonous stem reveals epidermis, cortex and stele -

Epidermis: It is protective in function and forms the outermost layer of the stem. It is a single layer of parenchymatous rectangular cells. The cells are compactly arranged without intercellular spaces. The outer walls of the epidermal cells have a layer called cuticle. The cuticle checks the transpiration. The cuticle is made up of a waxy substance known as cutin. Stomata may be present here and there. Epidermal cells are living. Chloroplasts are usually absent. A large number of multicellular hairs occur on the epidermis.

Cortex: Cortex lies below the epidermis. The cortex is differentiated into three zones. Below the epidermis, there are a few layers of collenchyma cells. This zone is called hypodermis. It gives mechanical strength to the stem. These cells are living and thickened at the corners. Inner to the hypodermis, a few layers of chlorenchyma cells are present with conspicuous inter cellular spaces. This region performs photosynthesis. Some resin ducts also occur here. The third zone is made up of parenchyma cells. These cells store food materials. The innermost layer of the cortex is called *endodermis*. The cells of this layer are barrel shaped and arranged compactly without intercellular spaces. Since, starch grains are abundant in these cells, this layer is also known as starch sheath. This layer is morphologically homologous to the endodermis found in the root. In most of the dicot stems, endodermis with casparian strips is not developed.

Stele: The central part of the stem inner to the endodermis is known as stele. It consists of pericycle, vascular bundles and pith. In dicot stem, vascular bundles are arranged in a ring around the pith. This type of stele is called eustele.

Pericycle: Pericycle is the layers of cells that occur between the endodermis and vascular bundles. In the sunflower stem, (*Helianthus*), a few layers of sclerenchyma cells occur in patches outside the phloem in each vascular bundle. This patch of sclerenchyma cells is called bundle cap or hardbast. The bundle caps and the parenchyma cells between them constitute the pericycle in the stem of sunflower.

Vascular bundles: The vascular bundles consist of xylem, phloem and cambium. Xylem and phloem in the stem occur together and form the vascular bundles. These vascular bundles are wedge shaped. They are arranged in the form of a ring. Each vascular bundle is conjoint, collateral, open and endarch.

Phloem: Primary phloem lies towards the periphery. It consists of protophloem and metaphloem. Phloem consists of sieve tubes, companion cells and phloem parenchyma. Phloem fibres are absent in the primary phloem. Phloem conducts organic food materials from the leaves to other parts of the plant body.

Cambium: Cambium consists of brick shaped and thin walled meristematic cells. It is two to three layers in thickness. These cells are capable of forming new cells during secondary growth.

Xylem: Xylem consists of xylem fibres, xylem parenchyma, vessels and tracheids. The xylem shows protoxylem and metaxylem elements. As protoxylem is present towards the centre and metaxylem towards periphery, the xylem endarch. Vessels are thick walled and arranged in a few rows. Xylem conducts water and minerals from the root to the other parts of the plant body.

Pith: The large central portion of the stem is called *pith*. It is composed of parenchyma cells with intercellular spaces. The pith is also known as medulla. The pith extends between the vascular bundles. These extensions of the pith between the vascular bundles are called primary pith rays or primary medullary rays. Function of the pith is storage of food.

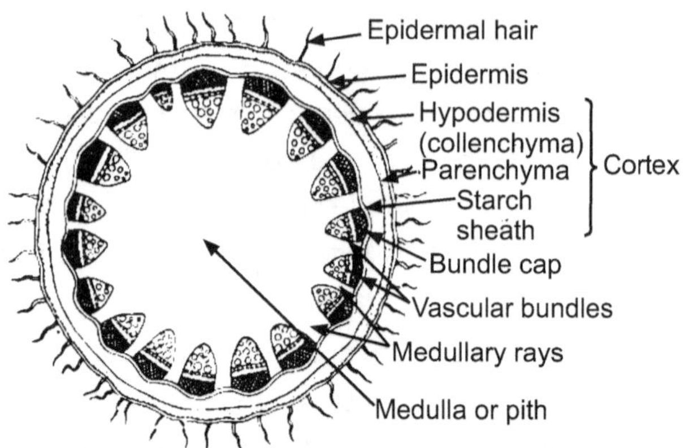

(a) Transverse section of Sunflower stem

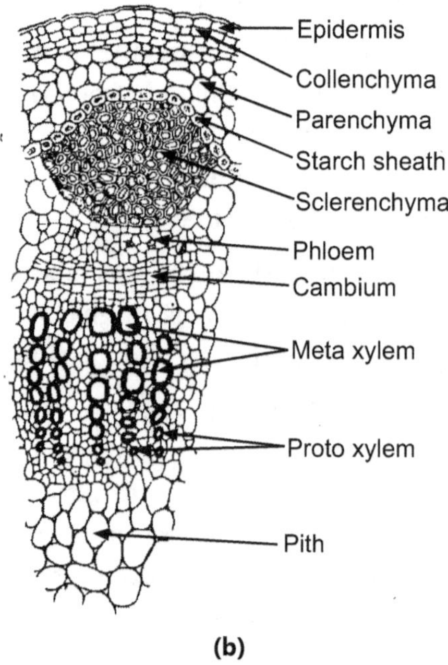

(b)

Fig. 3.5: Primary structure of dicotyledonous stem - Sunflower stem

3) Primary structure of a dicot leaf - Mango leaf (*Mangifera*)

Internal structure of dicotyledonous leaves reveals epidermis, mesophyll and vascular tissues.

Epidermis: A dicotyledonous leaf is generally dorsiventral. It has upper and lower epidermis. The epidermis is usually made up of a single layer of cells that are closely packed. The cuticle on the upper epidermis is thicker than that of lower epidermis. The minute openings found on the epidermis are called stomata. Stomata occur on the lower epdermis. A stoma is surrounded by a pair of bean shaped cells called guard cells. Each stoma opens into an air chamber. These guard cells contain chloroplasts, whereas other epidermal cells do not contain chloroplasts.

Mesophyll: The entire tissue between the upper and lower epidermis is called themesophyll. There are two regions in the mesophyll. They are palisade parenchyma and spongy parenchyma. Palisade parenchyma cells are seen beneath the upper epidermis. It consists of vertically elongated cylindrical cells in one or more layers. These cells are compactly arranged without intercellular spaces. Palisade parenchyma cells contain more chloroplasts than the spongy parenchyma cells. The function of palisade parenchyma is photosynthesis. Spongy parenchyma lies below the palisade parenchyma. Spongy cells are irregularly shaped. These cells are very loosely arranged with numerous air spaces. As compared to palisade cells, the spongy cells contain lesser number of chloroplasts. Spongy cells facilitate the exchange of gases with the help of air spaces. The air space that is found next to the stoma is called respiratory cavity or sub-stomatal cavity.

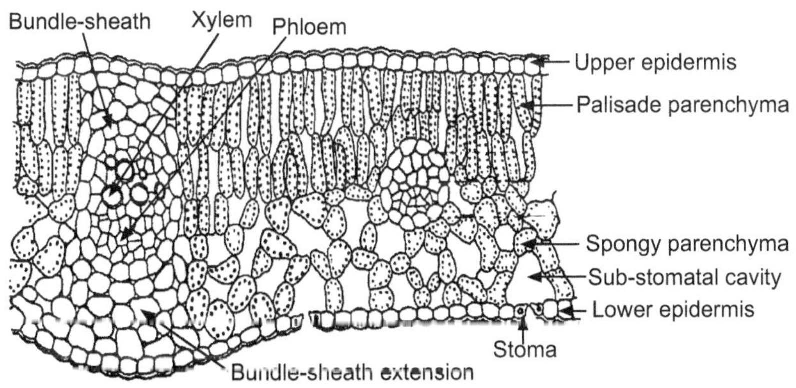

Fig. 3.6: Primary structure of a dicot leaf - Mango leaf

Vascular bundles: Vascular bundles are conjoint, collateral and closed. Xylem is present towards the upper epidermis, while the phloem towards the lower epidermis. Vascular bundles are surrounded by a compact layer of parenchymatous cells called bundle sheath or border parenchyma. Xylem consists of metaxylem vessels and protoxylem vessels. Phloem consists of sieve tubes, companion cells and phloem parenchyma.

3.2 NORMAL SECONDARY GROWTH IN DICOTYLEDONOUS ROOT AND STEM

3.2.1   Normal Secondary Growth in Dicotyledonous Root:

Increase in the girth or thickness of stem or root is known as secondary growth, which is generally shown by dicot stem and root. Whenever growth takes place by secondary tissues, it is referred as secondary growth. During secondary growth, secondary tissues are formed.

The dicotyledonous roots contain limited number of radial vascular bundles with exarch xylem. Usually pith is absent or very small. On the initiation of secondary growth, few parenchyma cells below each group of phloem become meristematic and forms cambium strips. These cambium strips divides to produce secondary tissues. The cells of pericycle present near protoxylem become meristematic and forms small cambium strips outside xylem strands. The first formed cambium strips present below phloem extends towards both the ends and joins with cambium strips formed from the pericycle, thus complete wavy ring of cambium is formed. The cambial cells produce more xylem elements than phloem. The first formed cambium produces secondary xylem much earlier, and the wavy cambium ring finally becomes circular. Now entire cambium ring becomes actively meristematic giving rise to secondary xylem on its inner side and secondary phloem towards outside. The secondary vascular tissues form a continuous cylinder and usually the primary xylem gets embedded in it. At this stage distinction can be made only by exarch primary xylem located in the centre. The primary phloem elements are generally seen in crushed condition. The cambial cells that originate from the pericycle lying against the groups of protoxylem function as ray initials and produce broad vascular rays. These vascular rays are also known as medullary rays. These rays are traversed in the xylem and phloem through cambium; this is characteristic feature of the roots.

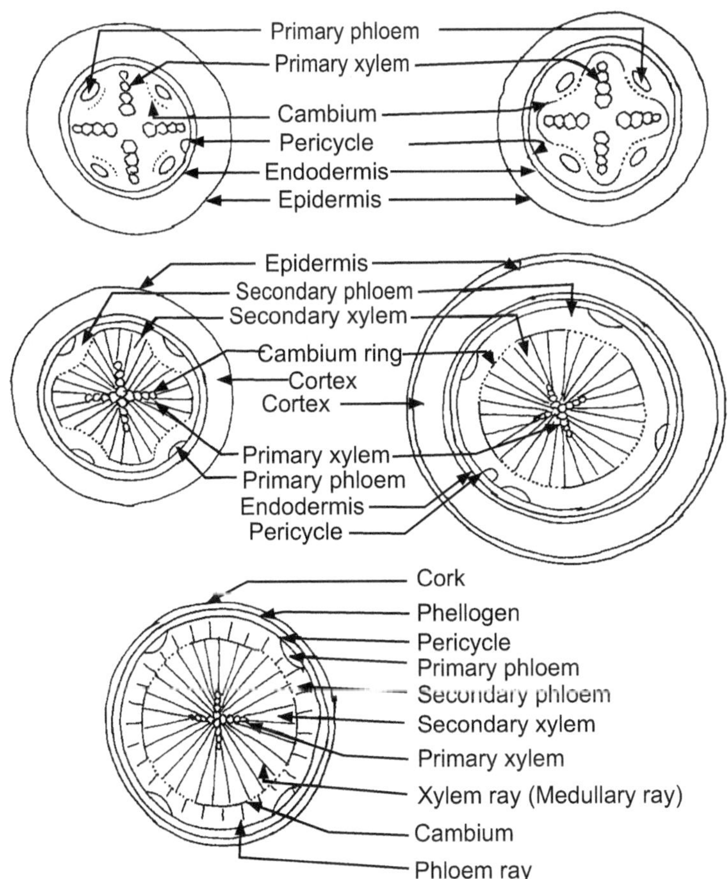

Fig. 3.7: Normal Secondary Growth in Dicotyledonous root

Simultaneously, a protective layer the periderm develops in the outer region of the root. The single layered pericycle becomes meristematic and divides, giving rise to cork cambium or phellogen. It produces a few brownish layers of cork or phellem towards outside, and the phellodorm on the inside. The pressure caused by secondary tissues ruptures the cortex with endodermis. Lenticels may also be formed due to secondary growth.

3.2.2 Normal Secondary Growth in Dicotyledonous Stem

Most of the dicotyledonous stem contain collateral and open end arch vascular bundles arranged in a ring. They are composed of xylem, phloem and cambium.

a) Cambium: The cambium is present in between xylem and phloem separating from each other known as vascular or fascicular cambium. The cells of cambium are meristematic. During the time of secondary growth, the parenchymatous cells of medullary rays lying between two vascular bundles become meristematic. Thus, new strips of meristematic cells are formed in between the vascular bundles known as interfascicular cambium. The interfascicular cambium connects the fascicular cambium and thus a complete ring of cambium is formed. The vascular cambium consists of two types of cells, the fusiform initials and the ray initials. The fusiform initials are vertically oriented and divide to form the elements of xylem and phloem. The cells of ray initials are smaller and isodiametric which give rise to vascular rays of parenchymatous cells.

b) Formation of Secondary Vascular Tissues: The cambium ring cuts-off new cells, both on outer and inner sides. The new cells formed on the outer side gradually modify into the elements of secondary phloem. The cells formed on the inner side gradually modify into secondary xylem.

Secondary Phloem: They consist of sieve tubes, companion cells, phloem parenchyma and phloem fibres. The primary phloem present on the outside gets crushed and is represented by small patches.

Secondary Xylem: Secondary xylem consists of vessels, tracheids, wood fibres and wood parenchyma. The vessels or trachea are most abundant in secondary xylem and are usually shorter than that of the primary xylem. The cambium ring forms more tissue on the inner side than on the outer side. As a result, secondary xylem forms the main bulk of the plant body and is generally called the wood. It's width increases with age. The primary xylem persists as conical projections towards the pith.

c) Vascular rays: Ray initials of the cambium ring form some narrow bands of parenchymatous cells. These cells extend radially from the pith to the phloem. These are called **medullary rays** or **vascular rays.** The rays present in xylem are xylem rays and the rays present in phloem are called **phloem rays.**

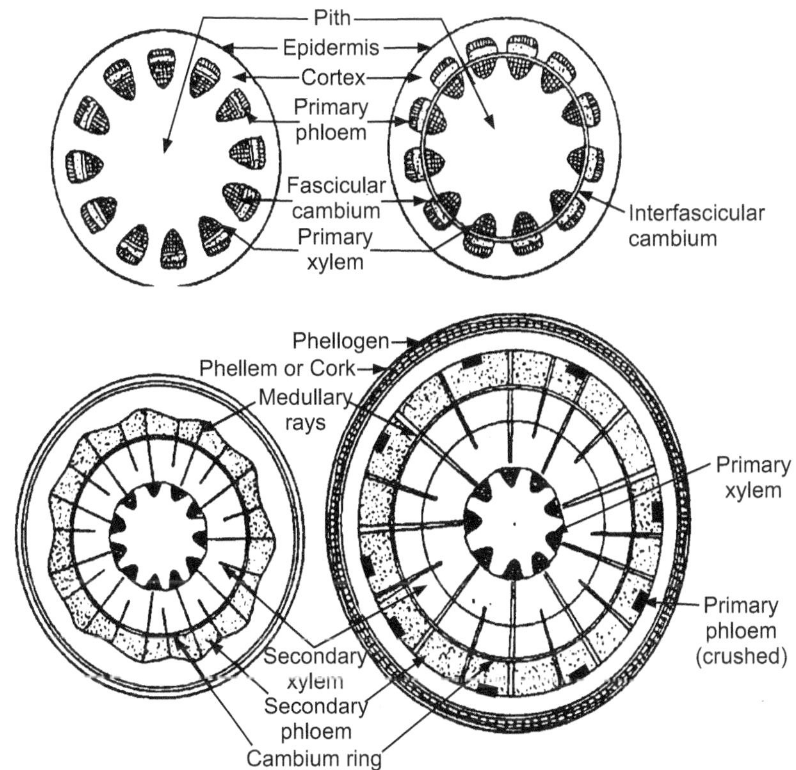

Fig. 3.8: Stages in the secondary growth of dicotyledonous stem (Diagrammatic)

d) Annual rings: The activity of vascular cambium is greatly affected by the variations in the climate. It is more pronounced in temperate regions. The cambium stops dividing in winter. In the spring season or early summer, the cambium becomes more active and produces a large number of vessels with wider lumen. These are called **spring wood** or **early wood.** During the autumn or winter season, the cambium becomes less active and produces vessels with narrow lumens. Tracheids and wood fibres are formed in large numbers. These woods are called **autumn wood** or **late wood.** Hence, the annual rings are formed year after year. In the oldest part of the tree, annual rings can be used in determining the age of a tree. In tropical regions, the climate is more or less uniform. Therefore, the annual rings are not well developed and does not correlate with the age of tree.

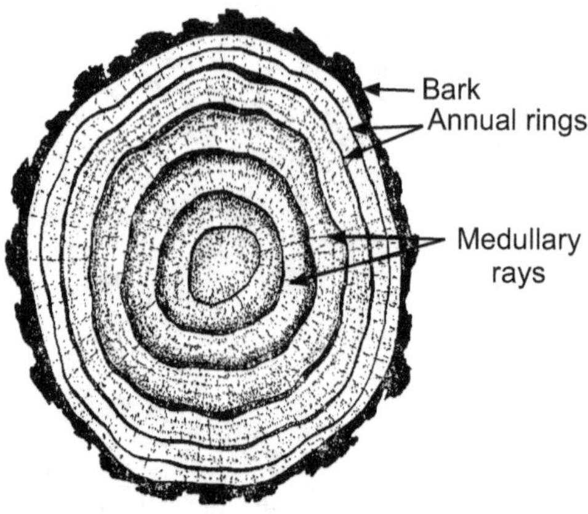

Fig. 3.9: T.S. woody stem showing annual rings

3.3 ANOMALOUS SECONDARY GROWTH IN *BIGNONIA* (DICOT.) AND *DRACAENA* (MONOCOT.) STEM

Anomalous secondary growth: When the secondary growth shows any deviation or change in the normal secondary growth it is referred as abnormal or anomalous secondary growth.

Anatomy of *Bignonia* (Dicot.) stem

Bignonia is a common garden woody climber that belongs to sub-class dicotylaedonae and family Bignoniaceae. The stem is green and quadrangular in the initial stage but after sufficient growth it becomes woody and circular in outline. At the initial stage of development the secondary growth is normal but after formation of some secondary tissues the cambium shows abnormal behavior. The Transverse section (T.S.) of old stem of *Bignonia* is more or less circular in outline. It shows the followings tissues:

1. **Epidermis:** It is outermost uniseriate layer of thin walled compactly arranged parenchyma cells. A thin layer of cuticle is present over it. The epidermis is ruptured here and there due to heavy secondary growth and formation of cork below it.

2. **Cork:** The epidermis is followed by cork layer which is composed of suberized cells.

3. **Cortex:** It consists of patches of collenchymatous cells, and parenchymatous cells.

4. **Endodermis:** It forms a distinct layer of barrel shaped cells.

5. **Pericycle:** It is parenchymatous with small isolated patches of sclerenchymatous fibres.

6. **Vascular System:** It consists of primary phloem, secondary phloem, cambium, secondary xylem and primary xylem.

   (a) Primary phloem arranged in small crushed patches and secondary phloem forms a complete ring, but at diagonal places it introduced into the secondary xylem to form four phloem furrows with sclerenchymatous bars.

   (b) Cambium is single layered, present in between xylem and phloem. It is depressed in the region of furrows.

   (c) Secondary xylem is in the form of ridges at four places due to phloem furrows and consists of tracheids, fibers, vessels and parenchyma.

   (d) The four furrows of phloem develop as a result of cambium produces more secondary phloem and less secondary xylem at four points. This is anomalous activity of the cambium.

   (e) Primary phloem and primary xylem are not visible because they are crushed due to secondary growth.

7. **Pith:** It is in the center of the axis and is composed of lignified parenchymatous cells.

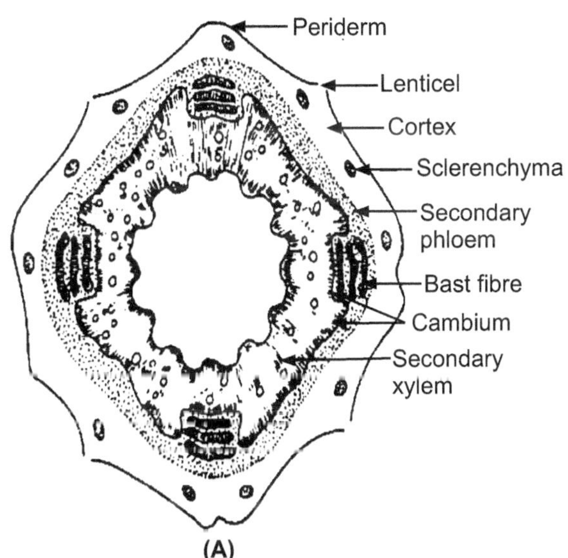

(A)

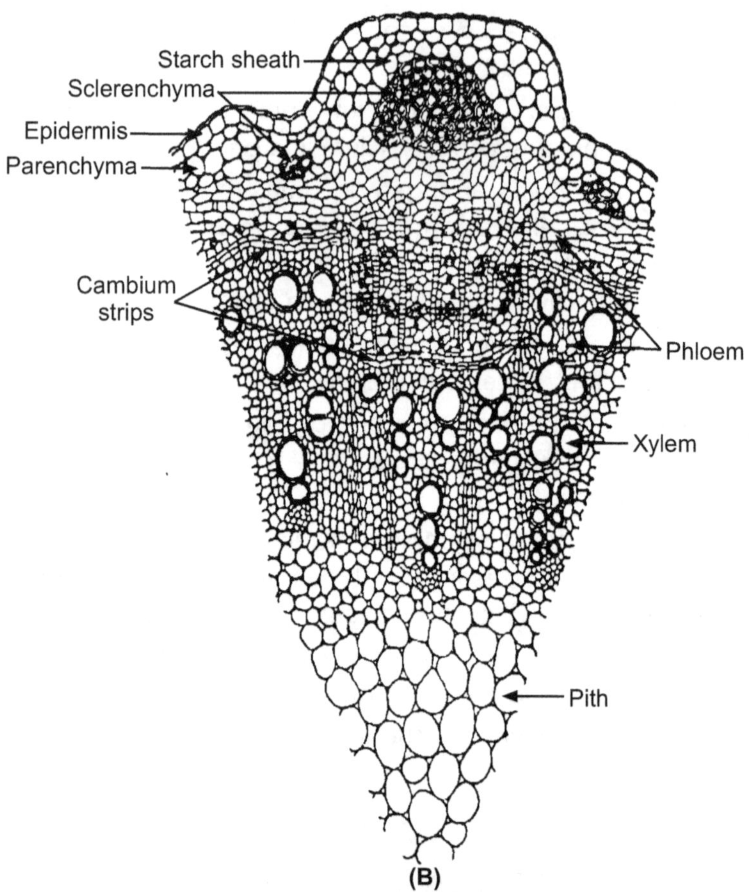

Fig. 3.10: T. S. of Bignonia stem (Dicot) showing Anomalous secondary growth – (A) Diagrammatic view (B) Enlarged cellular portion

The secondary growth in Bignonia is considered as abnormal for the following reasons:

In *Bignonia* the secondary growth is very peculiar and takes place due to the abnormal behavior of the normal cambium during later stages. In the beginning of the process, the cambium behavior is normal i.e. it produces equal amount of secondary xylem on the inner side and equal amount of secondary phloem on the outer side. Later on after producing some secondary tissues in normal manner, at four regions placed in crosswise manner, the cambium produces greater amount of secondary phloem on its outer side than the amount of secondary xylem on its inner side. Due to which wedges or patches of secondary phloem are formed at these four places which later on get embedded deeply into the secondary xylem. The production of unequal amount of secondary tissues i.e. less secondary xylem and more secondary phloem at four regions is abnormal behavior of the cambium ring. This abnormal activity is restricted to four crosswise placed strips which results into the production of xylem cylinder with four ridges and furrows or fan shaped xylem cylinder in old stem. The production of more amount of secondary phloem is useful to prevent the cracking or breaking of stem during strong winds and its climbing habit. As this abnormal growth is associated and beneficial with its habit it is called anomalous secondary growth.

Anomalous secondary growth in *Dracaena* (Monocot.) stem

Dracaena stem: *Dracaena* is the monocot plant which belongs to the family Dracaenaceae. The T. S. is circular in outline and shows the following structures: (Fig. 3.11)

(a) **Epidermis:** It is single layered, composed of rectangular cells and interrupted by the lenticels.

(b) **Ground tissues:** It is well developed and made up of parenchymatous cells.

(c) **Vascular bundle:**

  (i) Primary vascular bundles are conjoint, collateral, closed, few in number, larger in size and scattered in central region of ground tissues.

  (ii) During the secondary growth the massive cambium is developed from parenchyma outer side of the primary vascular bundle.

  (iii) Cambium tissue cuts off direct into many amphivasal or leptocentric secondary smaller vascular bundles.

  (iv) The secondary vascular bundles are different from the primary ones in the presence of small amount of phloem and in the absence of annular and metaxylem elements. The xylem is made up of only tracheids and small amount of xylem parenchyma.

(v) Secondary phloem of secondary vascular bundle consists of short sieve tubes and companion cells.

The secondary growth in *Dracaena* is considered as abnormal for the following reasons:

1. Monocots do not exhibit secondary growth but *Dracaena* shows secondary growth.

2. In normal secondary growth as in dicot, cambium produces secondary xylem towards the inner side and secondary phloem towards outer side. But in *Dracaena* all the secondary tissues are produced towards the inner side along with secondary xylem and secondary phloem cambium produces conjunctive tissues.

3. Primary vascular bundles are conjoint, collateral, closed but secondary vascular bundles are leptocentric.

4. Periderm is formed without formation of cork cambium.

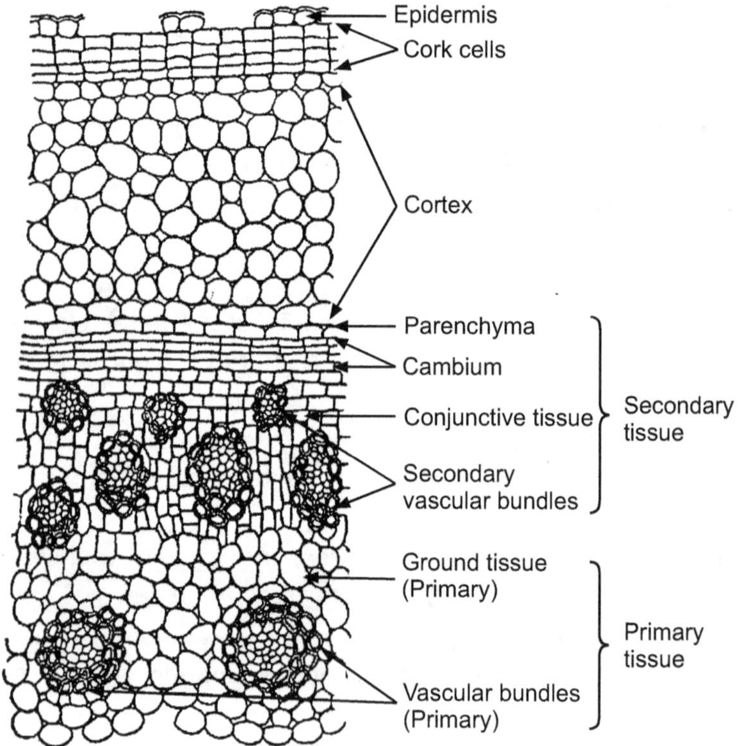

Fig. 3.11: T. S. of Dracaena stem (monocot) showing Anomalous secondary growth

5.   As this anomalous secondary growth is not associated with its morphology or physiological behavior its non-adaptive anomalous secondary growth.

3.4 PERIDERM AND LENTICEL

A) Periderm: The periderm is a protective layer formed in the older roots and stems of woody plants, when the outer tissues like the epidermis and some of the cortex rupture and die away after a considerable period of secondary growth. Leaves usually do not produce the periderm. Due to continuous secondary growth, the stem gradually keeps on increasing in width. In continuation of secondary growth the secondary tissues exert more pressure on phloem, cortical and epidermal cells and due to excess pressure, the phloem cells are pushed outwards and shape of cortical cells become flattened and finally epidermis get ruptured. Due to rupturing the living tissues are exposed to the external environment, at this stage the epidermis is replaced with secondary tissues which are protective in nature. The term periderm is applied to these tissues. The periderm is composed of three different types of tissues. i.e. i) Phellogen or cork cambium; ii) Phellem or cork and iii) Phelloderm or secondary cortex.

(i) Phellogen or Cork cambium: During the formation of secondary tissues, the pressure is caused from the internal tissues and the epidermis tends to rupture, a layer of phellogen or cork cambium is developed which is always meristematic. This meristematic layer originates immediately below the epidermis in the superficial layer of the cortex, the hypodermis. In transverse section the cells appear radially flattened rectangular in shape. In longitudinal section the cells appear rectangular, polygonal and very rarely irregular.

(ii) Phellem or cork: These cells are present outside the phelloderm and produce more quantity than phelloderm. As tissues of phellem are compactly arranged and with thick walls, due to this it becomes impervious to water and gases. They are rectangular in outline. When they are young, contain protoplast, but later on after maturation become dead and their walls become heavily suberized by fatty substances and lignified. Sometimes they may contain resins and tannins and become brown to yellow in colour.

(iii) Phelloderm: It is also known as secondary cortex. It is associated with storage of food material. They are composed of parenchyma, collenchyma and rarely of sclerenchyma cells. They are rectangular to irregular in outline and remain arranged in radial rows having small inter-cellular spaces among them.

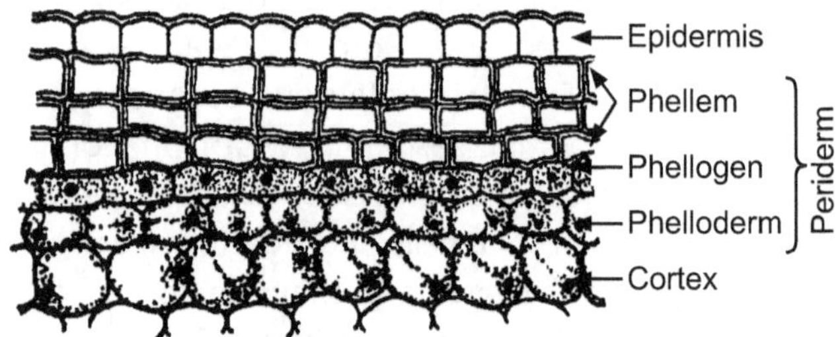

Fig. 3.12: Diagram showing Periderm formation

B) Lenticel:

1. Lenticels (means lens like cells) are pore like opening and look-like lens and present on the surface of the stem.

2. Lenticels are formed mostly in woody plants and especially where periderm is formed.

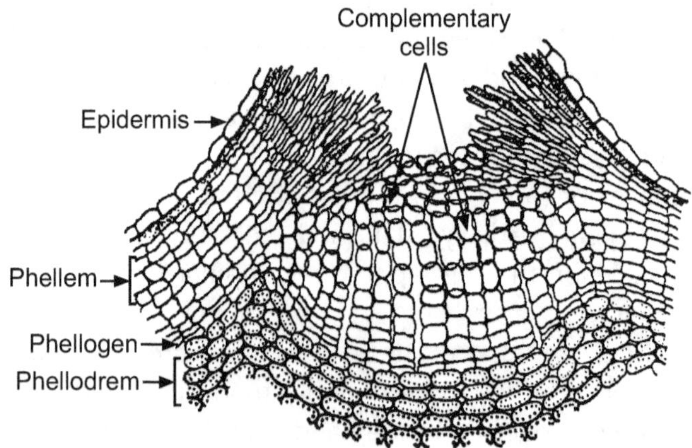

Fig. 3.13: Lenticel

3. Due to continuous secondary growth, the periderm is formed and the cork cells of the periderm are impervious to water and

gases. Thus, the gaseous exchange between the internal living cells and external atmosphere becomes difficult and therefore there is formation of lenticels.

4. Lenticels are useful for the gaseous exchange from internal living tissues to the external atmosphere.

QUESTIONS

A. Multiple Choice Questions:

1. The band like structure present in the radial and transverse walls of the endodermal cells are called ....
   - (a) Casparian strips
   - (b) xylem
   - (c) phloem
   - (d) pericycle

2. Radial type of vascular tissues are present in
   - (a) Mango leaf
   - (b) Maize root
   - (c) sunflower leaf
   - (d) Maize leaf

3. Scattered vascular bundles are present in.................
   - (a) Mango leaf
   - (b) Maize root
   - (c) sunflower leaf
   - (d) Maize stem

4. Vascular bundle surrounded by a sheath is called as...........................
   - (a) Bundle sheath
   - (b) Leaf sheath
   - (c) endodermis
   - (d) pericycle

5. In maize leaf ....patches of sclerenchyma are present above and below the large vascular bundles.
   - (a) one
   - (b) two
   - (c) three
   - (d) four

6. Tetarch vascular bundles are present in.....................
   - (a) sunflower root
   - (b) sunflower stem
   - (c) maize stem
   - (d) maize root

7. Increase in the girth or thickness of stem or root is known as.................
   - (a) medullary rays
   - (b) secondary growth
   - (c) primary growth
   - (d) None of these

8. When the secondary growth shows deviation in normal secondary growth it is known as.........................secondary growth.

   (a) abnormal   (b) normal

   (c) regular   (d) irregular

9. In Bignonia the secondary growth takes place due to the abnormal behavior of normal................

   (a) xylem   (b) phloem

   (c) cambium   (d) parenchyma

10. .... are pore like opening present on surface of the stem.

   (a) lenticels   (b) vascular bundles

   (c) casparian strips   (d) mesophyll cells

Answer key:

(1) – a, (2) – b, (3) – d, (4) – a, (5) – b, (6) – a, (7) – b, (8) – a, (9) – c, (10) – a.

B. Broad Questions:

1. Describe primary structure of Monocotyledonous root.
2. Describe normal secondary growth in dicotyledonous root
3. Describe normal secondary growth in dicotyledonous stem
4. Describe anomalous secondary growth in Bignonia stem
5. Describe anomalous secondary growth in Dracaena stem

C. Write Short Notes on:

1. Primary structure of Monocot leaf
2. Primary structure of Dicot leaf
3. Normal secondary growth
4. Abnormal /Anomalous secondary growth
5. Annual rings
6. Periderm
7. Lenticel

TISSUE SYSTEMS

4.1 EPIDERMAL TISSUE SYSTEM

The epidermal tissue system is the outer covering layer in a primary body which protects the inner tissues. It consists of cuticle, epidermis, stomata and epidermal outgrowths.

4.1.1 Cuticle

It is a layer of waxy substance known as cutin, on the outer wall of the epidermis. It is present on the epidermis of all aerial parts of plant body. Cutin layer means closely applied to the outer wall forming a continuous layer. It is probably secreted in liquid form by the epidermal cells and reaching outer wall, which becomes tough and hard. The cuticle is impermeable to water.

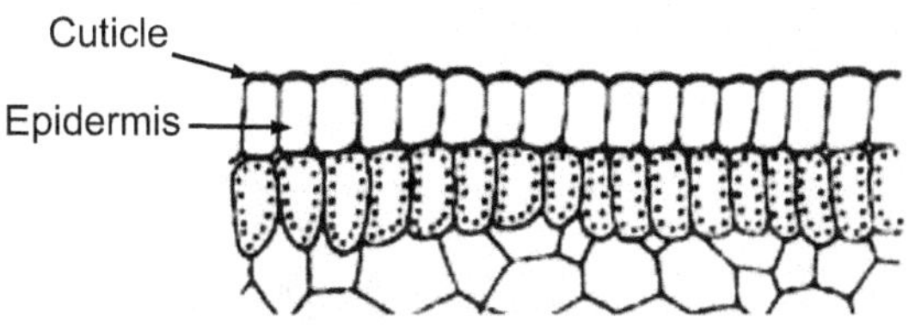

Fig. 4.1: Cuticle

Functions:

Prevent loss of water from inner cells. The cuticle is very thin in shady plants and those growing in moist plants. In xerophytes, the cuticle is thick. In some leaves, there is a coating of oil an epidermis e.g. *Nymphea*. It prevents wilting of leaf.

4.1.2 Epidermis

Epidermis is generally composed of single layer of parenchymatous cells compactly arranged without intercellular spaces. This word is derived from Greek word *epi-* upon, *derma-*skin. Epidermis consists of single layer of cells covering all organs of plant body, like stem, root, flower, leaf, fruit and seeds. It is continuous layer except small pores called stomata and lenticels.

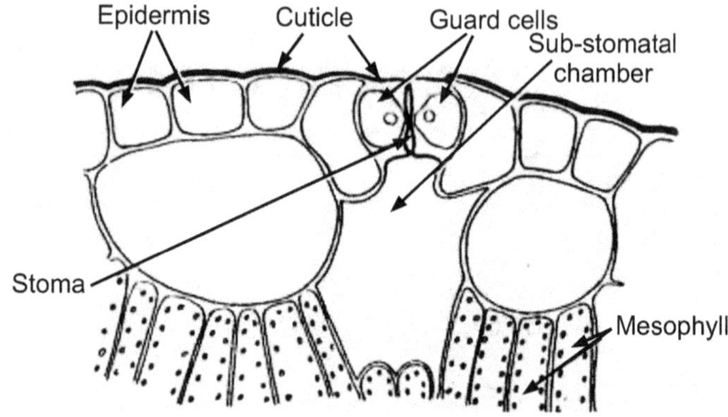

Fig. 4.2: Epidermis

Multiple Epidermis: Epidermis with many layers is called as multiple epidermis. Many layered epidermis is found in the leaves of plants like *Opuntia, Ficus,* and *Nerium.* It is made up of 3-5 layers of cells. In *Ficus* species multilayered epidermis prevents heating of the mesophyll cells below and helps to reduce transpiration.

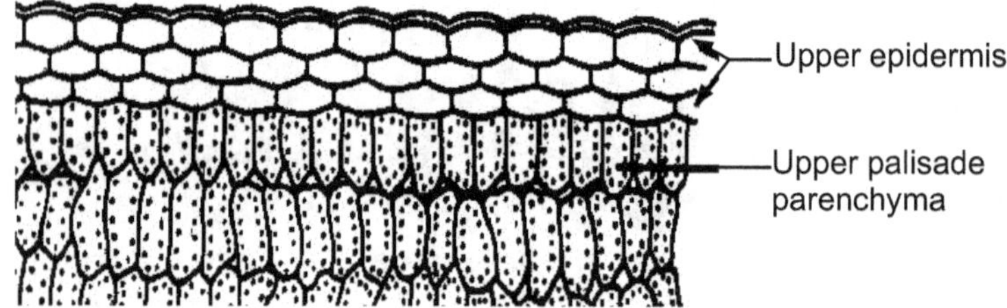

Fig. 4.3: Multiple Epidermis

4.1.3 Stomata

The stomata are minute pores which occur in the epidermis of plant. They are found on all aerial parts and are absent in roots. Stomata are made up of two guard cells, having an opening between them known as stomatal aperture. Generally the term stomata are applied to the stomatal opening and guard cells. There is a respiratory cavity below the stomata which allows gaseous exchange. The guard cells are living and contain chloroplast. Usually in the leaves of dicotyledons stomata are scattered whereas in the leaves of monocotyledons the stomata are in parallel rows, from this we can identify monocot and dicot plants. Number of stomata varies from few thousands to 100's of thousand per square cm or cm^2. Stomata occur on both upper and lower surface of leaf. But mostly they are found on lower surface. Stomata remain closed in night and it remain open during day time.

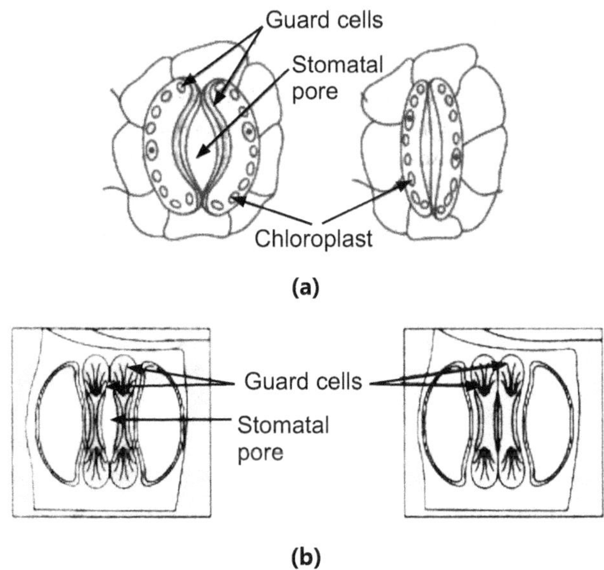

Fig. 4.4: (a) A dicot stomata and (b) A monocot stomata

Types of Stomata: The stomata are classified into five types on the basis of their morphology.

(a) Anomocytic / Ranunculous type: (Anomocytic – regular celled) Here stomata is surrounded by a limited number of subsidiary cells which are quite alike the remaining epidermal cells. The subsidiary cells are five in number.

(b) Anisocytic / Cruciferous type: (Anisocytic - unequal celled) Stomata is surrounded by three accessory cells of which one is distinctly smaller than the other two. e.g. *Petunia*.

(c) Paracytic / Rubiaceous type: (Paracytic - parallel celled) In this type stomata is usually surrounded by two subsidiary or accessory cells which are parallel to the long axis of the pore and guard cells. E.g. *Phaseolus*.

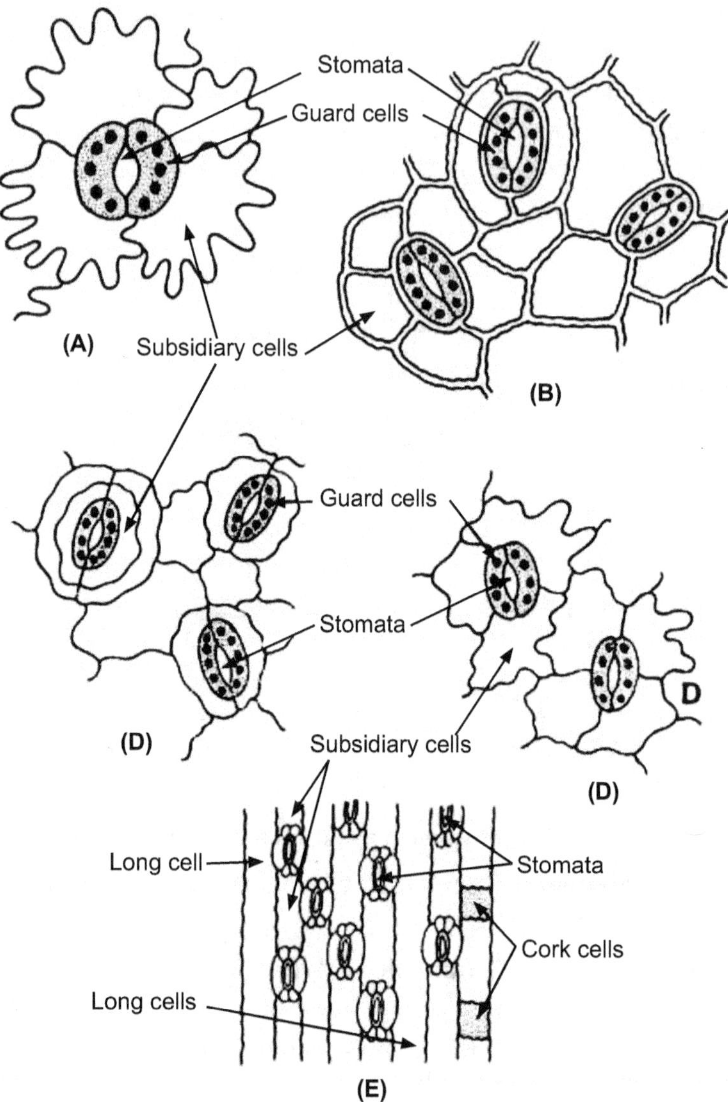

Fig. 4.5 : A. Anomocytic type; B. Anisocytic type; C. Paracytic type;
D. Diacytic type; E. Graminaceous type

(d) Diacytic / Caryophyllaceous type: (Diacytic - cross celled) Stomata is surrounded by a pair of subsidiary cells whose common wall is at right angles to the guard cells. e.g. Caryophyllaceae, Acanthaceae etc.

(e) Graminaceous type: This type of stomata is found in monocots like family Poaceae (Graminae). In this type of stomata guard cells are dumb bell shaped e.g. Cyperaceae, Poaceae (Graminae).

Functions:

1. They help in exchange of gases between the plant and atmosphere.

2. Evaporation of water also takes place through stomata.

4.1.4 Epidermal Outgrowths

In many plants, epidermal surface of the aerial organs show presence of different structures. These structures are considered as epidermal outgrowths. These outgrowths differ from thorns, stipules or spines because these structures are formed on epidermal surface only, while spines and other structures originate from the inner region of the organ. The epidermal outgrowths are generally called trichomes but this term is not correct because trichome means elongated filamentous structure while epidermal outgrowths except epidermal hairs are not like filament. Because of this reason the term trichome is to be restricted to epidermal hairs only. According to their morphological nature the epidermal outgrowths can be divided into five categories namely epidermal hair, peltate hairs, glandular hairs, water vesicles and root hairs.

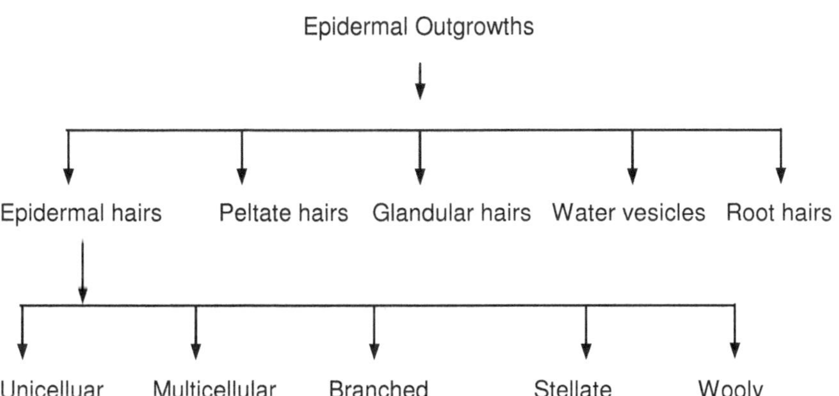

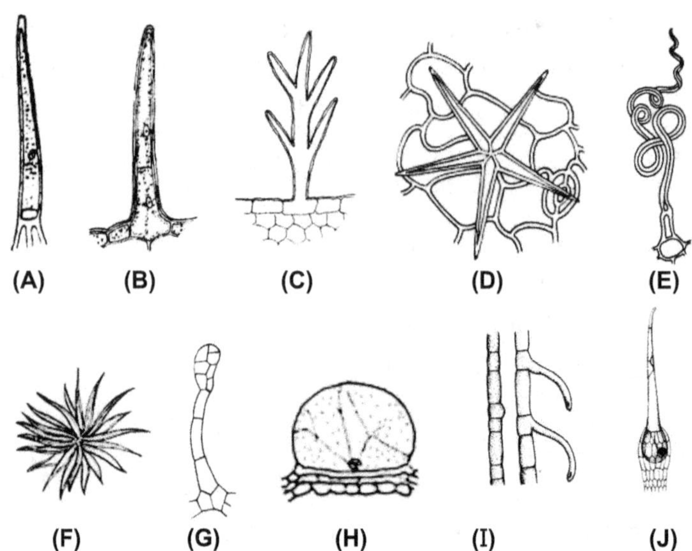

Fig. 4.6: A. Unicellular hair, B. Multicellular hair, C. Branched hair,
D. Stellate hair, E. Wooly hair, F. Peltate hair, G. Glandular hair,
H. Water vesicles, I. Root hair, J. Stinging hair

1.   Epidermal hairs: These hairs are very common outgrowths on stem and leaf surfaces. These structures are either unicellular or multicellular. When multicellular they are either unicliate of multiciliate. They either branched or unbranched. Unicellular or multicellular hairs which are unbranched can be observed in *Curcurbita* flower surface while branched can be observed in stem surface of many Amarantaceae members. In case of some plants, the upper part of epidermal hair is like star in such cases it is called stellate hairs. Common examples are *Pterospermum* of Sterculiaceae family. In some plants, this epidermal hair is very long soft delicate structure and this hair together form a wooly layer on the surface of organ from which they are produced. Common example is Cotton fibre which is of textile importance, is epidermal outgrowth from the outer seed coat. (Fig. 4.6 to E)

2.   Peltate hair: In some plant the epidermal outgrowth is in the form of flat plate like structure with or without stalk. In many plants between epidermis and plate few cells are present forming the foot of the hair. In some plants between foot and plate elongated stalk is present. Because of plate like upper part the entire structure is called peltate hair. (Fig. 4.6)

3. Glandular hair: In some plant, the surface of aerial organ is more or less sticky. It is mainly because certain substances or chemicals are secreted from the tip of epidermal hairs e.g. *Drosera* insectivorous plant. (Fig. 4.6)

In some plant, the chemicals which come out from such epidermal hairs after they break produce some irritation on skin, such hairs are called stinging hair e.g. *Urtica*. (Fig. 4.6) As the epidermal hair is in the form of small gland with long stalk it is called glandular hair.

4. Water vesicle: In some garden plants, the upper epidermal surface of the leaves show presence of small whitish beads like cells. These cells are full of water and look like small turgid balloon. Such cells water vesicles and their main function is storage of water e.g. *Mesembrythemum*. (Fig. 4.6)

5. Root hair: Root hairs differ from all other epidermal hair in one character i.e. all other hairs are additional structure on the epidermis while root hair is a part of epidermal cell. It is an elongation of the outer epidermal wall which is in direct contact of water supplying system. As compared to other epidermal hair root hair remains for very short duration because it is restricted to a particular zone of a root just behind the growing tip. (Fig. 4.6)

4.2 SECRETORY TISSUE SYSTEM

In plants, in addition to normal products certain substances are produced in the form of byproducts. Most of these byproducts are considered as secondary product because they are not utilized directly by the plants for the process of growth and development. Most of the products are useful to the plants in some other functions. e.g. Essential oil because of their volatile nature are responsible for peculiar smell which helps in pollination. When these oils are present in dead elements like xylem or wood, the region is protected from attack of fungi and bacteria. Another example is nectar attracts pollinating insects like bee which helps in pollination. In some plants such secondary products are secreted without their storage from certain regions of the plant body. In case of some plant some of these products are secreted but they remain accumulated or collected in certain special structure or cavities in the plant body. These structures or cavity they are released only after the structure breaks. Since in both the cases initially these secondary products are secreted most of them are called secretary products or substances. For the secretion of these substances certain specialized permanent tissues developed in plants. These permanent tissues are therefore secretory tissues. The secretory tissues are present either in the form of small patches in different region or they are present in the form of special structures or these patches of tissues as well as structures together formed a special system called secretory tissue systems. The secretory tissue system differs from other tissue system like epidermal tissue system, mechanical tissue system etc. Because of difference in the nature as well as in the structure the secretory tissues are classified into two categories –

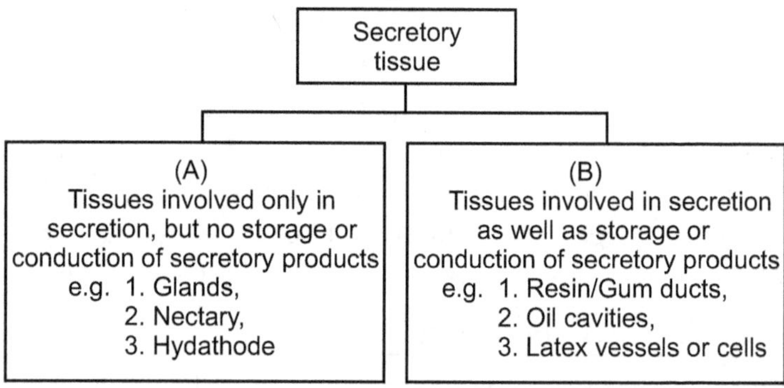

4.2.1 Tissues Involved Only in Secretion, but no Storage or Conduction of Secretory Products

1) Glands: (Fig. 4.7, B) Glands are the special structures present either on the outer surface or on the inner surface of certain aerial organs like leaves or modified leaves. Common examples are digestive glands in insectivorous plants. e.g. In *Drosera* on the surface of the leaves many glands are present. Each gland shows elongated stalk and club shaped head. In the outermost part of the head region secretory cells are present. These cells secrete the substances which collect on the surface of leaves in the form of shining dew drops. Due to which the plant is known as Sun Dew plant. This liquid includes many substances mainly carbohydrates and enzymes. Because of shining nature insects are attracted towards these dew drops and they caught on leaf surface with many hairs. In these hairs the insect gets entangled and then the action of enzyme begins due to which insect is killed. Then some of the glands present on the epidermal region of the leaf nitrogenous substances from the insect body are absorbed. Such digestive glands are also present on the inner surface of the pitcher in case of *Nepenthes*. The pitcher is modified leaf lamina and it is also an insectivorous plant.

2) Nectary: (Fig. 4.7) Nectar is an important sugary fluid secreted by some secretory tissues in the different regions of the plant. In some plants for secretion of nectar certain special structures are present and they are called nectaries or nectary. When a nectary is present in any part of an inflorescence or a flower it is called **floral nectary** and when it is present on any aerial part it is called **extra floral nectary**. Common example of floral nectary is *Euphorbia* cyathium inflorescence. In this inflorescence, inflorescence axis act as an involucral bracts together form a cup like structure which is normally dark green in colour.

On the outer surface of this cup distinct yellow coloured nectary is present. The vertical section passing through the nectary regions shows outermost layer consisting of columnar or elongated cells which are with dense cytoplasm and prominent nucleus. These cells are secretory cells specially developed for the secretion of nectar. In some flowers at the base of corolla tube surrounding the basal part of gynoecium nectaries are present. These nectaries secret the nectar which gets collected and remains at the base of corolla tube. To eat this nectar, pollinating agents like honey bees, small sun birds when visit the flower carry out pollination. It means in such plants the secretory product is for attracting the pollinating agents. Some plant like *Cassia* shows presence of extrafloral nectaries at the base of rachis especially very close to stem regions. The substances secreted by these nectaries are eaten by different types of ants which also carry out pollination in the species.

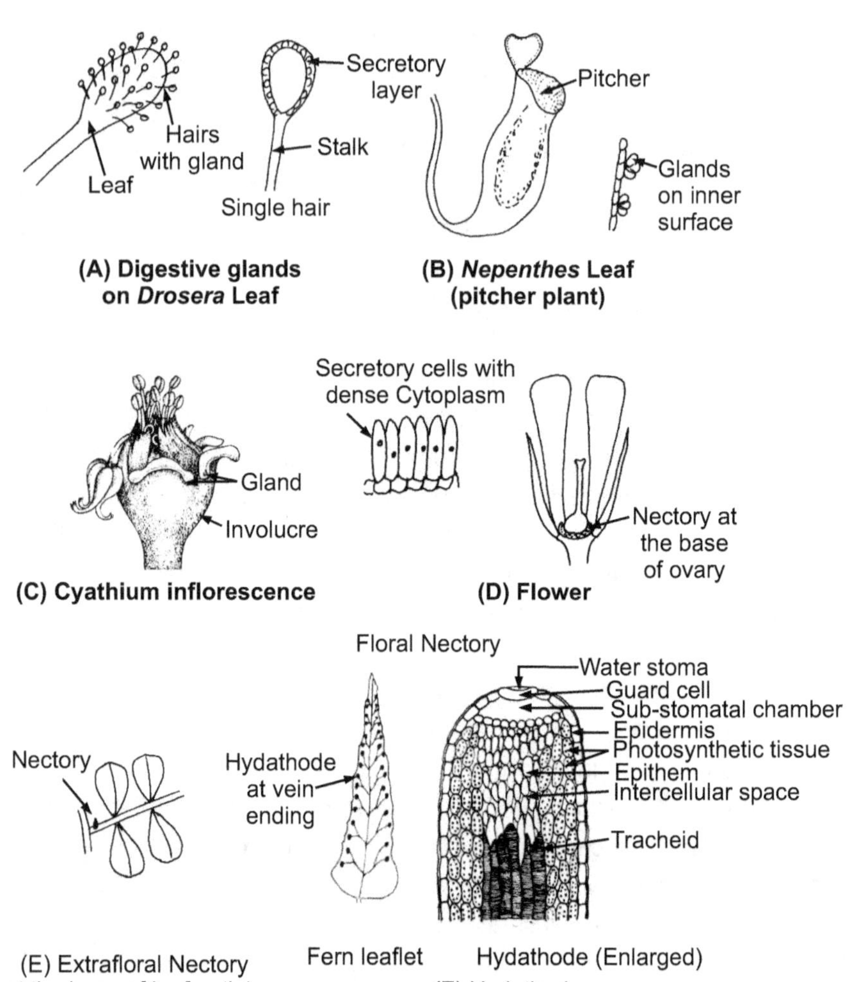

Fig. 4.7: Secretory Tissues involved only in secretion, but no storage or conduction of secretory products

3) Hydathode: (Fig. 4.7. F) In some herbaceous plant in the winter season during early morning hours water is lost from the leaf surface in the form of drops or droplets. When water is lost in the form of drops the process is called guttation. For the process of guttation in some plants special secretory structures are present and such structures are called hydathodes. In case of shade loving fern *Nephrolepis* such hydathodes are present in the marginal regions of a leaflet. Generally a hydathode is present at the end of vein or at veinlet ending. Each hydathode shows outermost layer epidermis with a distinct pore or opening at the tip. The pore is surrounded by normal epidermal cells called guard cells. Just below the guard cells a small cavity is present below the cavity at the tip of vein or veinlet ending a special tissue called epithem is present. The epithem tissue consists of elongated cells which are loosely arranged which means with many intercellular spaces. During night hours when there is no transpiration but there is sufficient absorption of water, root pressure increases and because of increased root pressure water from vein or veinlet ending is pushed in the epithem region. Because of same pressure from epithem excess water is pushed in the cavity just below the hydathode opening. Finally with cell pressure from the cavity, water is pushed in the outward direction which comes out through pore and gets collected in the form of water drop or droplets. Normally guttation occurs in herbaceous plants like many grasses and during winter season when transpiration is less and absorption is more. Water which is lost by guttation contains many minerals.

4.2.2 Tissues Involved in Secretion as Well as Storage or Conduction of Secretory Products

1) Resin/Gum Ducts – (Fig. 4.8) e.g. *Pinus, Thuja*

Resin is one of the important secondary products. Resins and oil resins in the wood protects the cells from adverse effects of low temperature, as well as from the attack of microorganisms. The resins are secreted and stored in sudden special structures called resin ducts. Initially the resin ducts are formed by separation of cells producing a small cavity and then the process continues in upward and downward direction producing an elongated space. With the elongation of cavity, small cells develop in the boundary region or in the periphery. These small cells are secretory in nature. The resin secreted by these marginal small cells get accumulated and later on conducted by an elongated structure called resin canals or resin ducts. Such ducts are very common in almost all aerial organs or gymnosperms. In some plants, gum formed by a special process called gummosis. It is also conducted through such structures. When only gum is stored the canals are called gum canals. In some plants mucilage a type of gum is stored and conducted. The canal is called mucilage canal or duct.

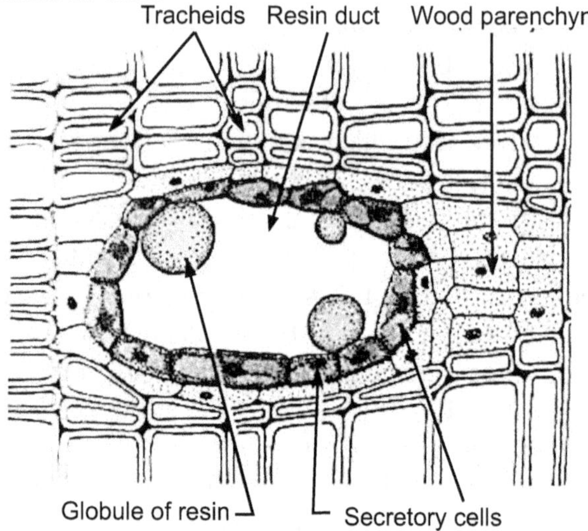

Fig. 4.8: Resin Duct

2) Oil Cavities: (Fig. 4.9. A):

Essential oil from herbaceous plants and trees are very important industrial product for their use in cosmetics, soap and incense industries. These oils are secreted, stored and conducted by special structures called oil cavities. The oil cavities are present in different parts in different plants. e.g. In case of *Cycas*, oil cavities are present mainly in leaves, In case of *Eucalyptus* oil cavities are mainly present in leaves while in case of coriander and other Apiaceae (Umbelliferae) plants they are present in all aerial organs. The oil cavities are formed by two different processes namely Schizogenous and Lysigenous. In case of Schizogenous cavities, the cavities developed by separation of cells and by increasing the size of intercellular spaces. In case of Lysigenous cavities, the cavity developed by this integration or lysis of some cells in a group. Normally in case of herbaceous plants Apiaceae (Umbelliferae) oil cavities are schizogenous while in case of trees like *Eucalyptus* and *Cycas* the oil cavities are lysigenous. Oil secreted by surrounding cell is accumulated and stored in oil cavities in the form of drop or droplets. In majority of plants such oils are yellowish in colour.

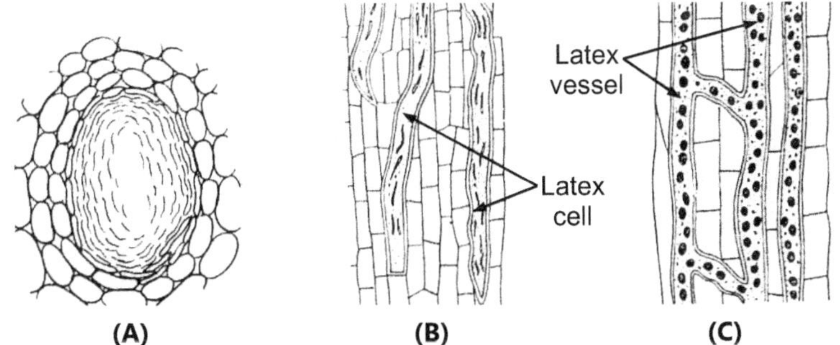

**Fig. 4.9: A. Oil Cavity of Citrus, B. non-articulate latex vessels,
C. articulate latex vessels**

3) Latex Vessels or Cells: (Fig. 4.9. B and C):

Like essential oils latex is also commercially important secondary product because it is utilized as a basic raw material in rubber industry. Latex is a thick fluid containing large number of different substances mainly carbohydrates, proteins, oils and with some volatile substances. These substances are secreted as well as conducted in certain special structures called latex cells or vessels. Cells are the structures which are long tubular but without branching while vessels are also tubular structures but with many branches and interconnections between the different regions. Laticiferous ducts may be i) non-articulate latex ducts or latex cells and ii) Articulate latex ducts or latex vessels. Both latex cells and vessels originate from promeristem right from embryonic stage which means it is a genetical character but the presence of cells or vessels is not characteristics of any family. It means in the same family some plants show latex cells while some genera are with latex vessels.

Important plants producing latex includes families like Euphorbiaceae, Papavaraceae, Caricaceae, Apocynaceae, Moraceae etc. *Heavea branziliensis* is important latex yielding plant economically important for rubber.

4.3 MECHANICAL TISSUE SYSTEM

The group of cells or tissue which performs the function of mechanical support are called as mechanical tissues. Various organs of plant body are subject to mechanical injury from the forces of nature. Stems of herbaceous plants are subject to horizontal pressure on one side and compression on the other due to high winds. Woody trees have to bear the weights of massive branches and the crown of leaves. The roots are subject to radial pressure and are pulled up when the stem is shaken by high velocity of winds. Leaves are subject to pressure at right angles to the surface and the lamina is likely to be torn by winds. Massive branches are subject to bending pressure due to their own weight. To guard against their various injuries, plants have developed a mechanical tissue system. Mechanical tissue system is made up of different types of cells viz. collenchyma, sclerenchyma, xylem tracheids, vessels, fibres, sclereids.

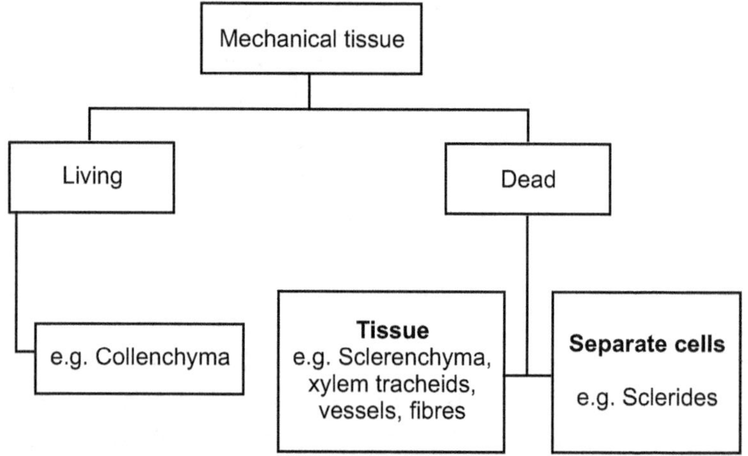

Collenchyma and sclerenchyma cells give maximum mechanical strength. Collenchyma walls get thickened at the corners due to deposition of pectin, cellulose, hemicellulose and protein. The collenchyma cells are also elastic due to presence of hydrated pectin on the walls.

In case of some tall tress like gymnosperms the mechanical support is provided by trachieds in the vascular bundles and conducting regions as well as by sclerides which develops either in the form of individual cells or in the form of small patches. In gymnosperms sclerenchyma fibres are either few or absent and therefore tracheids in the wood regions carry out conduction of water as well as additional function of mechanical support. Sclerides are the sclerenchyma cells with relatively very less length. These cells are also dead cells and show wall with very thick deposition of lignin.

Because of very thick secondary wall and short structure even a single cell can give sufficient mechanical support to the organ in which it is produced. Because of this character of sclerides in almost all higher plants sclerides are produced in almost all organs when sufficient mechanical support is required. In many plants therefore sclerides are produced in petioles, pedicels as well as in fruit wall. From mechanical support point of view sclerenchyma and sclerides are two important tissues developed in plants. These mechanical tissues produced in different organs from root to fruit together form the mechanical tissue system. During formation of mechanical tissue system, different types of mechanical tissues in different plant organs are distributed in appropriate manner considering different types of stresses to which the plant organs are exposed. While distributing these mechanical tissues in the different organs one basic principle is followed by engineers during the construction of buildings and bridges. This basic principle is maximum support with minimum expenditure of material.

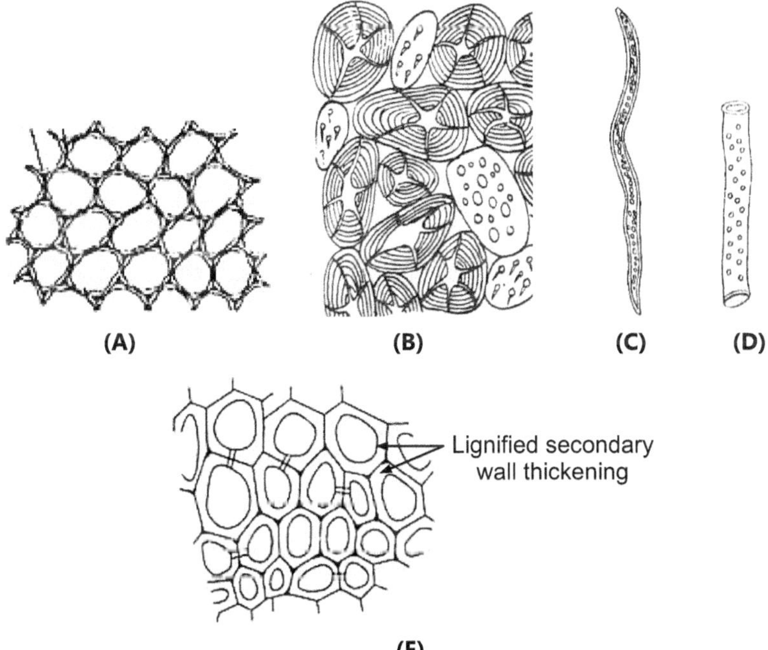

Fig. 4.10. A. Collenchyma, B. Sclerides, C. Tracheid,
D. Vessel, E. Sclerenchyma

1. **Distribution of Mechanical tissues:** Mechanical tissues are developed to resist different strains and stresses. In case of majority of land plants, the different organs require distribution of mechanical tissues mainly for four different types of nature.
    a. Distribution for inflexibility (Flexible nature)
    b. Distribution for inextensibility (Extensive nature)
    c. Distribution for incompressibility (Compressible nature)
    d. Distribution for shearing stresses.

(a) Distribution for inflexibility (Flexible nature):

In case of **Sunflower** stem mechanical tissues are present in the hypodermal region as well as in the pericyclic region. These patches of mechanical tissues are distributed in such a way that two patches in peripheral region lie on the same radius but at opposite direction. These patches of mechanical tissues together form the patches of I-girders, while the central parenchyma in the pith and cortex together form web in the I-girder.

All these I-girders meet each other exactly in the center of stem where all radii meet because of this reason these I girders forms a composite system and stem shows distribution of mechanical tissues in the form of composite I-girders.

In case of **Maize** stem the mechanical tissues are also in the form of I-girders but the system includes sclerenchyma in the hypodermal region in the form of patches and ground tissue with scattered vascular bundles in the form of I-girders forming web region.

In case of stem of **Lamiaceae** member's i.e. squarish stem the girders are diagonal eye girders. The four patches of collenchyma in the four corners of hypodermis and pericyclic sclerenchyma patches together form the flanges while the cortical pith region of parenchyma forms the web.

These different examples indicate that for inflexibility in aerial organs like herbaceous stem distribution of mechanical tissues is in the form of eye girders places in different directions and position as per nature or organ of stem.

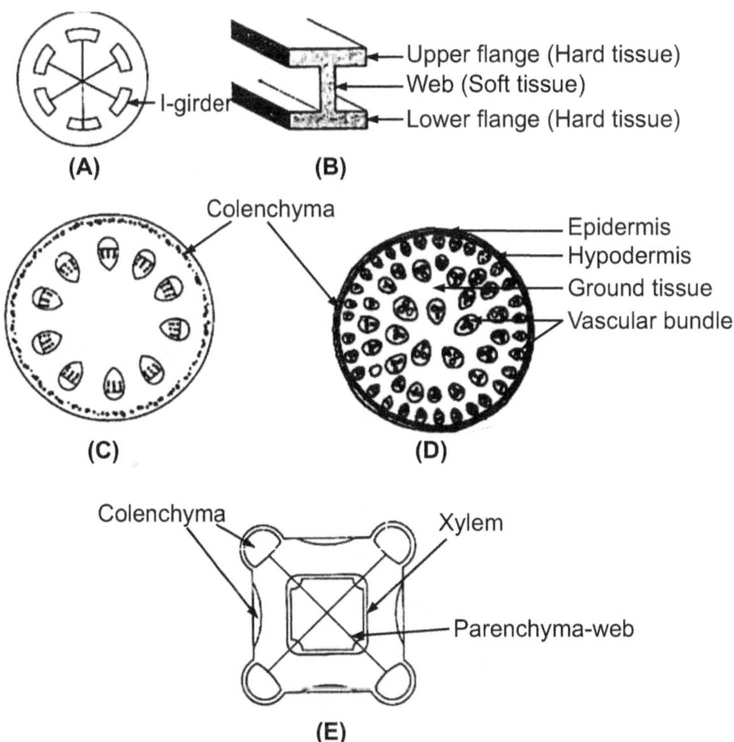

Fig. 4.11: Composite I-girders in different plant – A. A composite I-girder, B. Portion of I-girder, C.Mechanical tissue sytem in Sunflower stem, D. Mechancial tissue system in Maize stem, E. Mechanical tissue system in Lamiaceae family

(b) Distribution for inextensibility (Extensive nature):

For inextensibility the organs requires mechanical tissues in central regions generally root system which grows towards the gravity in the side fixes the entire vertical system in the soil. When vertical system is exposed to stresses from different reactions e.g. Wind currents or air currents. The root system gets pulled in the upward direction. To avoid uprooting of plant system and to keep plant anchored or fixed in the soil the main root show mechanical tissues in the form of ring in pericycle region. In addition to this in some roots, mechanical tissues are present surrounding the radial vascular bundle patches. These mechanical tissues are generally jointed or united forming a thick wall in that region. Since this thick wall is with soft tissue i.e. pith parenchyma in the centre it is in the form of cylinder. Such cylinder of mechanical tissue in the central part gives sufficient extensible nature to the root system and it avoids uprooting under adverse conditions. In case of some plants having relatively week but erect stem form aerial organs additional roots are produced e.g. Maize, Jowar, and trees like Banyan. These additional roots are either in the form of stilt root or in the form of prop roots. Because of their aerial origin such roots in addition to centrally placed mechanical tissues shows additional ring of mechanical tissues generally sclerenchyma in the hypodermal regions. These roots as well as additional mechanical tissues exhibit flexible nature and plant necessary support to remain erect.

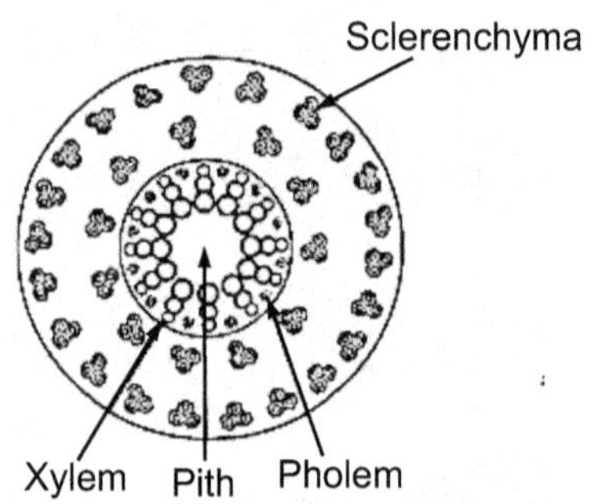

**Fig. 4.12: Inextensibility: Patches of Sclerenchyma
forming cylinder (with pith)**

(c) Distribution for incompressibility (Compressible nature)

In case of large trees after few years of growth, when many branches with large number of leaves produced and relatively thick crown is formed at the upper end of vertical stem. Because of weight of this crown the vertical stem is pushed downward direction which means exposed to compression. Under such conditions the stem requires compressible nature and mechanical tissues for incompressibility to bear this weight at one end. For this reason the secondary growth begins the secondary xylem tissue is added in the relatively greater amount in the central region and this secondary xylem forms the central rod called as wood. Because of this wood the vertical stem can tolerate the weight of large crown.

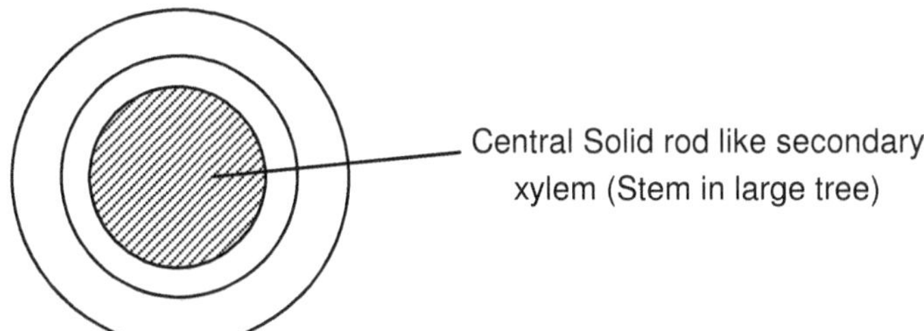

Fig. 4.13: Incompressibility: Stem in large tree (Diagramatic)

(d) Distribution for Shearing stresses:

In case of leaves in most of the large trees the mechanical tissues are distributed without any definite direction or arrangement. In leaves the veins form complete mesh or net because of reticulate arrangement in vein on either side or completely surrounded mechanical tissues are present. Because of different directions of veins these mechanical tissues are also placed in different regions and in different directions. The leaves when exposed to wind currents simultaneously form different directions the lamina remains without tearing i.e. intact therefore to tolerate shearing stress in case of dicot plants the leaves shows reticulate venation.

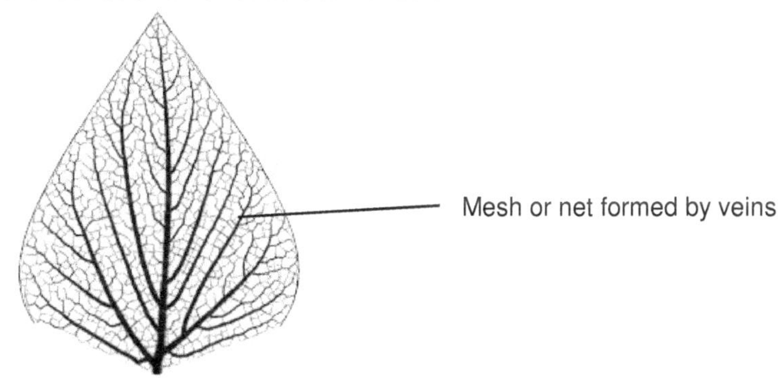

Fig. 4.14: Shearing stress in leaf

QUESTIONS

A. Multiple Choice Questions:

1. Multiple epidermis is present in ....leaf

    (a) Nerium   (b) Mango

    (c) Sugarcane   (d) Maize

2. In...................type stomata is usually surrounded by two subsidiary cells which are parallel to long axis of pore and guard cells.

   (a) Anomocytic   (b) Paracytic

   (c) Anisocytic   (d) Diacytic

3. ....shaped guard cells are present in Poaceae

   (a) Bean   (b) Spherical

   (c) Dumbbell   (d) rectancular

4. In Drosera...........................type of hairs are present on the leaf.

   (a) Unicellular   (b) Glandular

   (c) wooly   (d) peltate

5. Floral nectary is present in cyathium inflorescence of......................

   (a) Euphorbia   (b) Urtica

   (c) Nepenthes   (d) Cotton

6. In Nephrolepis...................are present in the marginal regions of leaflet.

   (a) Digestive glands   (b) Wooly hairs

   (c) Hydathodes   (d) Stellate hairs

7. ....cavity is developed by lysis of some cells in a group.

   (a) Lysigenous   b) Schizogenous

   (c) stomatal   (d) mesophyll

8. ....is an important latex yielding plant

   (a) Maize   b) Surgarcane

   (c) Hevea   (d) Citrus

9. In .... the mechanical tissues are in the form of I-girders.

   (a) Maize stem   (b) Surgarcane leaf

   (c) Citrus fruit   (d) Mango leaf

10. In ....plant stilt roots are present for giving additional mechanical strength.

   (a) Citrus   (b) Nephrolepis

   (c) Maize   (d) Mango

> **Answer key:**
>
> (1) – a, (2) – b, (3) – c, (4) – b, (5) – a, (6) – c, (7) – a, (8) – c, (9) – a, (10) – c.

B. Long Answer Type Questions:

1. Describe in short epidermal tissue system

2. Describe epidermal outgrowths in plants

3. Describe in brief secretory tissue system

4. Describe in brief mechanical tissue system

5. Describe distribution of mechanical tissues in plants

C. Short Notes:

1. Stomata

2. Epidermal hairs

3. Nectary

4. Resin ducts

5. Distribution of mechanical tissue for inflexibility

6. Distribution of mechanical tissue for inextensibility

1
CHAPTER

ENZYMES

1.1 INTRODCUTION

Enzymes are the special types of proteins which accelerate the rate of biochemical reactions taking place in the living organisms. It is clear that all the enzymes are proteins but not all the proteins are enzymes.

Frederick Kuhne (1878) for the first time coined the term "enzyme" for these special proteins. In (1926) J. Summer isolated and purified enzyme urease in crystalline form for the first time. The enzymes act as organic catalyst in biochemical reactions. Their catalytic property depend upon the integrity of their own protein structure. The enzymes got this catalytic capability due to the presence of some special properties.

Definition: An enzyme can be defined as "A complex chemical protein produced in a living cell which catalyses a biochemical reaction".

In other words "enzymes are proteinaceous biocatalysts which accelerate the rate of reaction". (En = in, zyme = leaven or living). Enzymes are very specific about the substrate and type of the reaction they catalyse. They are required in very low concentration and remain unchanged at the end of the reaction. So far about 2000 enzymes are discovered.

1.2 CLASSIFICATION AND NOMENCLATURE OF ENZYMES

Classification: International Union of Biochemistry (IUB) has given a detail classification of enzymes. IUB has given certain rules and regulations for the classification and nomenclature of enzymes (1961). The enzymes are generally named by adding a suffix 'ase' to the name of the substrate. e.g. Urea-urease, phosphate-phosphatase.

Enzymes can be classified by following ways:

1. The reaction catalysed.

2. Presence or absence at a given time.

3. The regulation of action.

4. The place of action.

1.2.1 Classification of Enzymes Based Upon the Reaction They Catalysed

Enzymes are broadly divided into six groups based on the type of reaction catalysed.

Table 1.1

Sr. No.	Class	No. of sub-classes	Reaction	Example
1	Oxidoreductases	97	Oxidation-Reduction	Dehydrogenase
2	Transferases	09	Transfer of functional group	Phosphatases
3	Hydrolyses	13	Hydrolytic reactions	Enzymes acting on peptide bonds e.g. pepsin, amylase
4	Lyases	99	Addition or loss of double bonds	Aldolases
5	Isomerases	55	Isomerization of substrate	Epimerase
6	Ligases	06	Union of two molecules	Glutamine synthetase

1.2.2 Classification Based on Presence or Absence at a Given Time

(a) **Inducible enzymes:** Those enzymes that are synthesized by the cell whenever they are required. Synthesis of these enzymes requires an inducer. e.g. invertase, p galactosidase.

(b) **Constitutive enzymes:** These are constantly present in the cell irrespective of inducers. e.g. enzymes of glycolysis.

1.2.3 Classification Based on the Regulation of Enzyme Action

(a) **Regulatory enzymes:** The action of these enzymes is regulated depending upon the status of the cell. The action of regulatory enzymes is either increased or decreased by the modulator at a site other than active site, called the "allosteric site". e.g. phosphofructokinase, glutamate dehydrogenase.

(b) **Non-regulatory enzymes:** The action of these enzymes is not regulated. e.g. succinate dehydrogenase.

1.2.4 Classification Based on the Place of Action

(a) **Intracellular enzymes:** These enzymes are produced by the cell and act inside the same cell are known as intracellular enzymes. e.g. enzymes of glycolysis.

(b) **Extracellular enzymes:** These enzymes are produced by a cell but act outside that cell. e.g. trypsin.

(1) Details of classification based upon the reaction catalysed is described as below:

Classes:

(i) Oxidoreductases: The enzymes of this group are catalysing oxidation-reduction reactions. In these reactions one compound is oxidised and other is being reduced after the completion of enzyme action. These enzymes act by transferring the electrons and hydrogen ions. e.g. oxidases, reductases, dehydrogenases.

(a) Glutamate dehydrogenase:

$$\text{Glutamic acid} + \text{NAD} \underset{\text{enzyme}}{\rightleftharpoons} \alpha - \text{ketoglutarate} + NH_3 + \text{NADH}$$

(b) Lactate dehydrogenase:

$$\text{Lactate} + NAD^+ \underset{\text{enzyme}}{\rightleftharpoons} \text{Pyruvate} + \text{NADH}$$

(c) NAD Oxidoreductase:

$$\text{Alcohol} + NAD^+ \underset{\text{enzyme}}{\rightleftharpoons} \text{Aldehyde} + \text{NADH} + H^+$$

(d) Glyceraldehyde 3 phosphate dehydrogenase:

$$\text{Phosphate} + \text{NAD} \underset{\text{enzyme}}{\rightleftharpoons} 1.3 \text{ bis-phosphoglycerate}$$

(2) Transferases:

These enzymes catalyse the reactions where functional groups or group of atoms are transfered from one molecule to other. During these reactions various functional groups like carboxyl, formyl, methyl, acetyl and phosphate are transfered from donor molecule to other acceptor molecule. e.g. hexokinases, transaminases, kinases etc.

(a) Glutamate-pyruvate amino transferase:

$$\text{Glutamate + Pyruvate} \xrightleftharpoons{\text{enzyme}} 2-\text{oxyglutarate + alanine}$$

(b) Glucokinase:

$$\text{ATP + D-glucose} \xrightleftharpoons{\text{enzyme}} \text{ADP + o-glucose-6-phosphate}$$

(3) Hydrolyses:

This group also has large number of enzymes. These enzymes are also called as special transferases. These enzymes participate in the reactions where hydrolysis of compounds takes place. In these reactions the donor group is transfered to water molecule. These enzymes cause addition of water to a variety of bonds like C–O, C–N, C–C, O–P and C–S bonds. They cause the cleavage of complex substrate molecules e.g. lipases, peptidases, phosphatases, amidases, ribonucleases, proteases, carbohydrases, esterases etc.

(a) ATPase:

$$\text{ATP} + H_2O \xrightleftharpoons{\text{enzyme}} \text{ADP + iP}$$

(b) Lipase:

$$\text{Lipid} + H_2O \xrightleftharpoons{\text{enzyme}} \text{Fatty acid + Glycerols}$$

(c) Sucrose invertase:

$$C_{12}H_{22}O_{11} \xrightleftharpoons{\text{enzyme}} C_6H_{12}O_6 + C_6H_{12}O_{12}$$

$$\text{Sucrose} \qquad\qquad \text{Glucose} \qquad \text{Fructose}$$

(4) Lyases:

These enzymes catalyse the breakage of a compound into two substances by mechanism other than addition of water (hydrolysis) or oxidation or reduction. These enzymes act on C – C, C – O, C – N,

C – S and C – halide bonds. The resulting product always has a double bond. e.g. decarboxylases, aldolases etc.

(a) Decarboxylase: removes CO_2 from keto acid.

(b) Fructose, diphosphate aldolase:

$$\text{Fructose} - 1 - 6 - \text{diphosphate} \xrightleftharpoons{\text{enzyme}} \text{Glyceraldehyde} - \text{3-phosphate (3PGAL)} + \text{Dihydroxy acetone phosphate (DHAP)}$$

(5) Isomerases:

These enzymes catalyse the isomerization type of reactions. In these reactions the structural rearrangements of atoms in the substrate molecule takes place. The enzymes of this group are named on the basis of type of isomerism like isomerases, epimerases, mutases etc. They may include cis-trans, keto-enol and aldose-ketose etc.

(a) Epimerase:

$$\text{D} - \text{xylulose} - 5 - \text{phosphate} \xrightleftharpoons{\text{enzyme}} \text{D} - \text{ribulose} - 5 - \text{phosphate}$$

(b) Triose-phosphate isomerase:

$$\text{D} - \text{glyceraldehyde} - 3 - \text{phosphate} \xrightleftharpoons{\text{enzyme}} \text{Dihydroxy acetone phosphate}$$

(c) Phosphohexose isomerase:

$$\text{Glucose} - 6 - \text{phosphate} \xrightleftharpoons{\text{enzyme}} \text{Fructose} - 6 - \text{phosphate}$$

(6) Ligases-(or synthetases):

These enzymes are also known as synthetases. These enzymes catalyze the reactions in which union of two compounds takes place. These reactions require energy. In these reactions pycophosphate bond of ATP is broken down and linkage between two molecules takes place. These enzymes form C – O, C – S, C – N and C – C bonds. e.g. acetyl co-A, Carboxylases, DNA ligase, synthetase and DNA polymerase etc.

(a) Glutamine synthetase:

$$\text{Glutamate} + NH_3 + \text{ATP} \xrightleftharpoons{\text{enzyme}} \text{Glutamine} + \text{ADP} + \text{iP} + H_2O$$

(b) Pyruvate Carboxylase:

$$\text{Pyruvate} + CO_2 + ATP \underset{\text{enzyme}}{\rightleftharpoons} \text{Oxaloacetate} + ADP + iP$$

(c) Aspargine Synthetase:

$$\text{Aspartic acid} + NH_2 + ATP \longrightarrow \text{Aspargine} + AMP + iP$$

Nomenclature of enzymes:

Naming or nomenclature of the enzymes is an important step in the study of enzymes. Nomenclature of the enzyme is done by the method given by IUB (International Union of Biochemistry, 1961). According to IUB enzymes are classified into six major classes. These six classes are made on the basis of general type of chemical reaction they catalyse. Usually suffix 'ase' is attached to the name of substrate. The important features of IUB system are given below:

(1) The enzymes are classified into six major classes on the basis of general types of reaction they catalyse.

(2) Each class is further divided into several sub-classes on various basis like-type of bond split or formed, chemical group removed or transferred.

(3) Main classes and sub-classes are indicated by index numbers.

(4) The third digit indicates the sub-sub-classes.

(5) The forth digit indicates the systematic specific name of the enzyme.

On the basis of this method all the enzymes are named by four digit E. C. Code. The IUB has given the following major classes, sub-classes, sub-sub-classes and specific names.

Major Classes:

1. Oxidoreductases
2. Transferases
3. Hydrolases
4. Lyses
5. Isomerases
6. Ligases

Sub-classes:

1.1 Oxidoreductase, acting on the CH.OH group of the donor.

2.1 Transferase, transfering one carbon atom.

3.1 Hydrolase, which act on ester links.

4.1 Lyase, action on C – C bond.

5.1 Isomerase, which acts as racemase and epimerase.

6.1 Ligase, which forms C-O bonds.

Sub-sub-classes:

1.1.1 Oxidoreductase, acting on the CH.OH group of donor, with co-enzyme NAD or NADP as the acceptor.

2.1.1 Transferase, transfering one carbon groups and a methyl transferase.

3.1.1 Hydrolase, action on carboxylic ester links.

4.1.1 Lyase, C – C lyase, carboxylase.

5.1.1 Isomerase, acting as a racemase and epimerase on amino acids and derivatives.

6.1.1 Ligase, which forms C – O bonds, amino acid – RNA ligase.

e.g. Alcohol dehydrogenase is named as 1.1.1.1.

1. major class – oxidoreductase

1.1 oxidoreductase, acting on the CH.OH group of donor.

1.1.1 oxidoreductase, acting on the CH.OH group of the donor, with co-enzyme NAD or NADP as the acceptor.

1.1.1.1 Alcohol NAD: oxidoreductase (or Alcohol dehydrogenase).

Specific or trivial names:

1.1.1.1 alcohol dehydrogenase

2.1.1.1 nicotinamide methyl transferase

3.1.1.1 carboxylestorase

4.1.1.1 pyruvate decarboxylase

5.1.1.1 alanine racemase

6.1.1.1 tyrosyl-s-RNA synthatase

4.1.3.1 iso-citrate dehydrogenase

1.1.1.27 lactate dehydrogenase

2.7.1.1 hexokinase

1.3 STRUCTURE AND PROPERTIES OF ENZYMES

Structure:

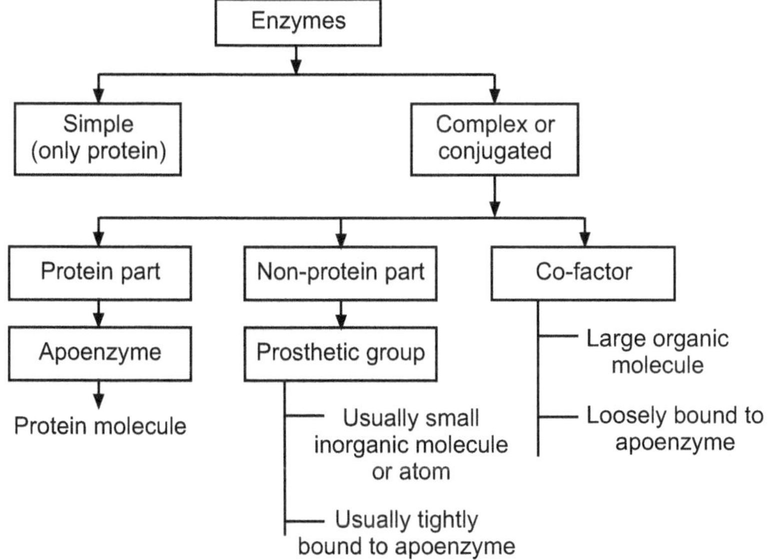

Fig. 1.1

Enzymes are protein molecules. They are made up of amino acids. They may contain 100 to 1000 amino acids. These amino acids are joined together in a long chain, which is folded to produce a three dimensional structure (3D).

The shape of an enzyme is very important because it has a direct effect on how it catalyse a reaction.

An enzymes shape is determined by the sequence of amino acids in its structure, and the bonds which form between the atoms of those molecules. Different types of enzymes have different shapes and functions because the order and type of amino acids in their structure is different.

Enzymes are very specific about which reactions they catalyse. Only molecules with exactly the right shape will bind to the enzyme and react. These are the 'reactants or substrate' molecules. The part of the enzyme to which the reactant binds is called as the 'active site'.

As the enzymes are proteins they are active only when they acquire their proper conformation. (amino acid sequence).

Enzymes show primary, secondary, tertiary, and quaternary structural folding. The (a) Primary, (b) Secondary, (c) Tertiary and (d) Quaternary structures are described as below:

(a) Primary structure:

- Ala – Glu – Val – Thr – Asp – Pro – Gly –

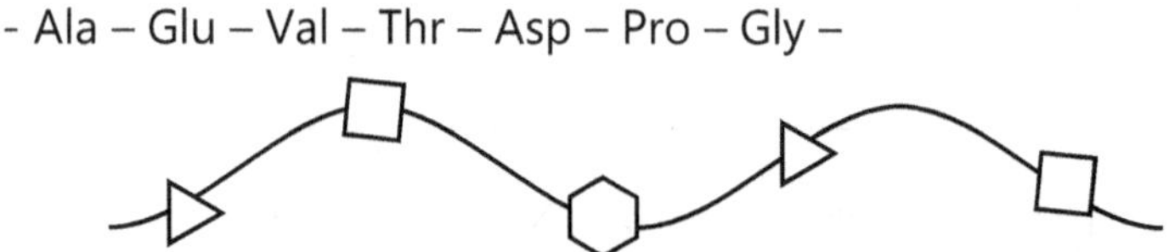

Fig. 1.2: Simple chain of amino acids

The primary structure of the enzyme show only a long polypeptide chain with many amino acids in a sequence. In this structure more than 30 amino acids in peptides and from 200 to 1000 amino acids in proteins are joined together by peptide bonds (or amide bonds). These amino acids are in between the carboxyl and amino terminals (ends). They are called as N-terminus and C-terminus.

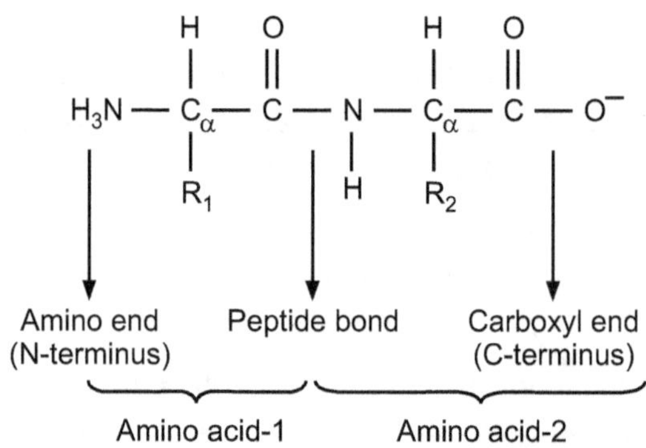

Fig. 1.3: Molecular structure of simple enzyme

The skeleton of all proteins consist of a [N - C_α (R) – C (o)–] repetitive unit. Only R-group side-chains vary. The sequence of proteins should be written from left to right i.e. From the N-terminus to C-terminus. e.g. yeast protein contains 466 amino acids. The average molecular weight of an amino acid is 113 daltons (Da).

(b) Secondary structure:

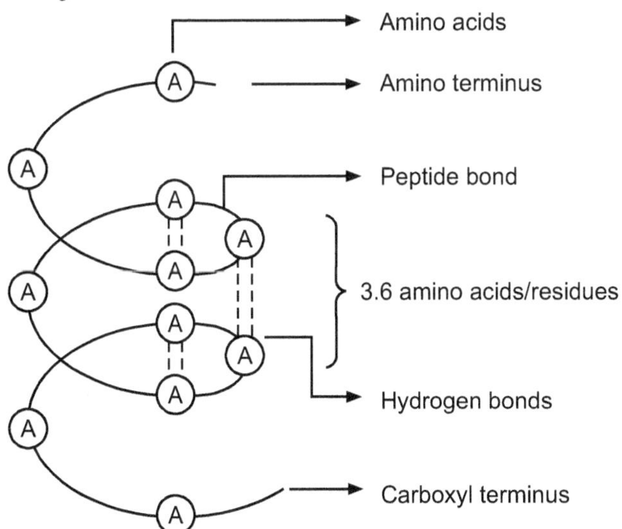

Fig. 1.4 : α - helix structure of enzymes

In the secondary structure the long polypeptide chain show specific folds. These folds gives the polypeptide chain a helical or spiral structure called as α-helix. Each turn shows 3.6 amino acids or residues, which means that there is one residue every 100° of rotation (360/3.6°). The vertical distance of each turn of helix is 1.5 A° and the distance between two amino acid is 5.4 A° (angstroms). The specific foldings are supported and stabilised by the hydrogen bonds. These hydrogen bonds are formed between two amino acids. These amino acids are present in the same chain. The α-helix structure shows two-end amino terminus and carboxyl terminus. Like α-helix structure secondary proteins also show β-sheets, β-turns (hairpins or loops) structures which are more complicated arrangements of amino acids. The two terminals play important role in biosynthesis of enzymes.

(c) Tertiary structure (3D structure):

Tertiary structure of the enzymes show the folded 3 Dimensional (3D) structure of a protein. It is also known as the 'native structure' or

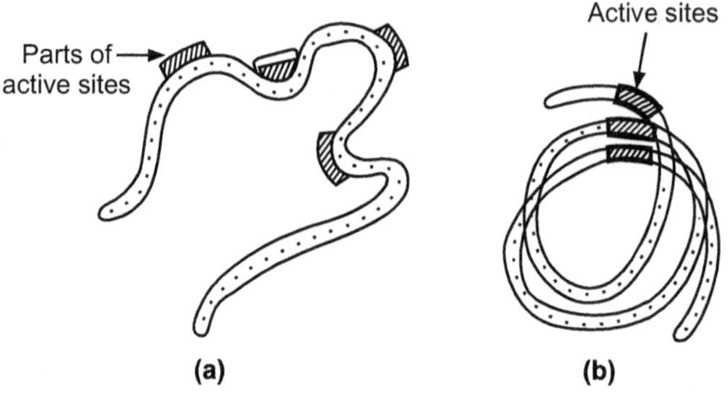

Fig. 1.5

'active conformation'. In the tertiary folding the active sites are created by bringing together the active groups. During the process the folded regions and random regions of a protein come very close to each other. This three dimensional structure is maintained by the disulphide bonds and hydrogen bonds. As the parts of active sites are coming close to each other various types of interactions like ionic and hydrophobic interactions begins. The Van-der-Waals forces acting on the protein make the 3D structure compact and efficient. The tertiary structure is responsible in bringing the amino acids very close to each other. This causes the development of active sites with specific amino acids on the surface of enzyme molecule.

The development of 3D structure and active sites are responsible for the biocatalytic activity and action specificity of enzymes. The active sites are the regions with specific amino acids with their side chains left open to form bonds with specific substrates.

The 3D structure of enzymes remain functional under favourable conditions of temperature, pH etc. If the temperature and pH is increased beyond optimum limits the 3D structure of enzyme is disrupted and enzymes looses its biocatalytic property.

As most of the enzymes are proteins, but some RNA molecules are also working as enzymes. Such RNA molecules have catalytic properties and catalytic activities. Therefore they should be considered as enzymes.

(d) Quaternary structure (conjugated proteins):

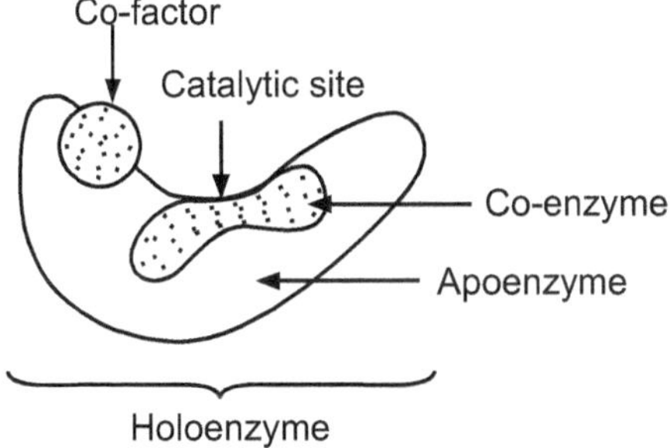

Fig. 1.6: General structure of conjugated enzyme

Holoenzyme = Apoenzyme + Prosthetic group

The conjugated enzyme shows two sub-units.

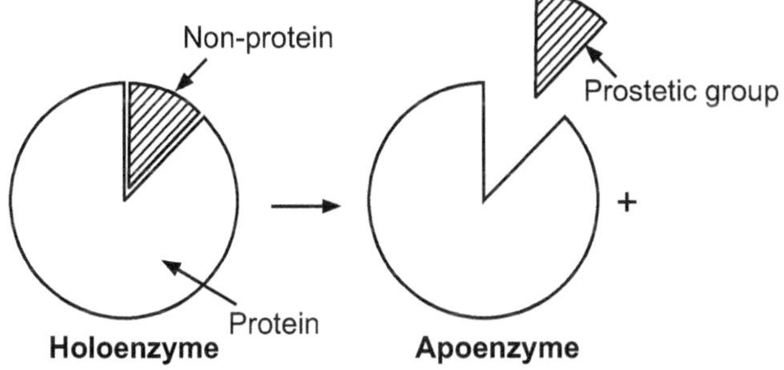

Fig. 1.7: Components of conjugated enzyme

The conjugated protein enzymes are formed by two sub-units. A protein part and a non-protein part. The protein part is called as apoenzyme (e.g. Flavoprotein) and the non-protein part is called as 'prosthetic group'. The complete conjugate enzyme, consisting of an apoenzyme and the prosthetic group is called as 'holoenzyme'. The enzyme remain active only when the apoenzyme and prosthetic group remain attached or linked, to each other. When these sub-units are separated, the enzyme looses catalytic property. The apoenzyme or prosthetic group alone cannot catalyse any reaction.

The nature of prosthetic group is variable. It may be organic or inorganic (metal ion) in nature. The organic prosthetic group are called 'co-enzymes' and the inorganic prosthetic groups are called 'co-factors'.

(e) Co-enzymes, co-factors and Iso-enzymes:

Co-enzymes (Co-meaning assist, help, co-operative):

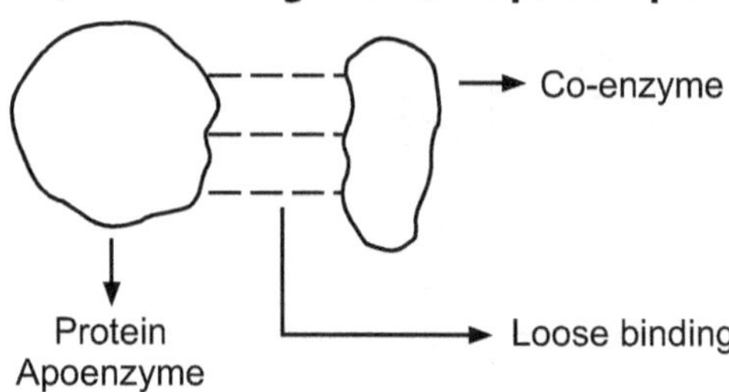

Fig. 1.8: Co-enzymes

The meaning of co-enzymes is the additional factor which helps for the action of enzymes. They are non-proteinaceous. Their relationship with the enzymes is temporary and established as per the need of the occasion. They are chemically different, heat stable, low molecular weight organic compounds, and required in very small amount. They are biosynthetically similar to the vitamins.

There are two categories of co-enzymes:

(i)　Those which are really part of the active center. They assist the protein side chains in the catalysis. e.g. pyridoxal phosphate.

(ii) The second category of co-enzymes are called as co-substrates. e.g. NAD^+.

The roles of co-enzymes are:

(i)　They act as co-enzymes or second substrate.

　　e.g. pyruvate + NADH $\longrightarrow$ Lactate + NAD^+.

　　Here NADH act as a co-enzyme or second substrate.

(ii) They help in transferring of groups either hydrogen or groups other than hydrogen.

(iii) Specific activity of a co-enzyme is the number of units of co-enzyme present in one milligram of enzyme protein.

Following are some of the important co-enzymes:

(a) **Nicotinamide Adenine Dinucleotide or NAD:** It is needed in different reactions for transfer of terminal hydrogen from substrate. It may convert to NADH.

(b) Nicotinamide Adenine Dinucleotide Phosphate or NADP: It act as a hydrogen transporter. It is very similar to NAD.

(c) Flavine Nucleotides (FMN and FAD): Flavine Mono Nucleotide (FMN) also act as co-enzymes of dehydrogenases often called flavoproteins or yellow ferments. They are used for transfer of hydrogens from substrate or as an intermediate electron acceptor by specific dehydrogenases.

(d) Ferro-porphyrins: They associate with cytochromes. They are divalent and trivalent.

(e) Lipoic acid: They act as hydrogen transporter in oxidative decarboxylation reactions.

(f) Coenzyme-A (Aceylation co-enzyme): They are useful in oxidation of fatty acids. They also act as substrate.

(g) Tetrahydrofolic acid (FH_4): It is the active form of a vitamin or folic acid.

(h) S-Adenosyl Methionine: It is a transporter of methyl group. It acts as a substrate for methylation reactions.

(i) Biotin: It is used in carboxylation reactions.

(j) Pyridoxal Phospate: This derives from pyridoxine or vitamin B_6. It is a part of active site of various enzymes of amino acid metabolism.

(k) Cobamide Co-enzymes: These are derived from cobalamine or vitamin B_{12}.

Co-factors:

These are non-proteinaceous inorganic elements attached to the protein part of enzymes. They are generally the ions of different inorganic elements. They are also called metal activators. These are tightly bound with the enzyme proteins and called as prosthetic groups. The co-factors show some important properties:

(i) They have the ability to form various types of chelates.

(ii) Some co-factor ions have an empty electron orbit which acts as an electron sink.

Some of the important co-factors are:

(a) Cu^{++}: Prosthetic of cytochrome oxidase.

(b) Fe^{++} or Fe^{+++}: Prosthetic group of catalase, cytochrome oxidase.

(c) Mn^{++}: Co-factor of orginase and other co-enzymes.

(d) Mg^{++}: Co-factor of polymerases.

(e) Mo^{++}: Co-factor of xanthine-oxidase and Nitrate Reductase.

(f) Ni^{++}: Co-factor urease.

(g) Ca^{++}: Co-factor for ATPase.

(h) Zn^{++}: Co-factor for dehydrognease, DNA polymerase, Carbonic unhydrase.

Isoenzymes (Isozymes) Marker and Moller (1959):

The isoenzymes are multiple forms of an enzyme performing on the same substrate and giving the same end product. They only have difference in the amino acid sequence. The isozymes of an enzyme differ in their electrophoretic mobility, isoelectric pH and binding with some chelates and ligands. They are the products of different genes. They are also subjected to different kinds of regulation. Isozymes can be separated by sedimentation, electrophoresis, chromatography, gel filtration etc.

Lactic acid dehydrogenase is an ideal example of isozyme formation. It catalyses the oxidation of lactic acid to pyruvic acid with the help of NAD^+.

$$\text{Lactic acid} + NAD^+ \underset{}{\overset{\text{Lactic acid dehyrogenase}}{\rightleftharpoons}} \text{Pyruvic acid} + NADH^+ + H^+$$

There are various forms of lactic acid dehydrogenase as LDH1, LDH2, LDH3, LDH4, LDH5. Each type is a tetramer with four polypeptide subunit.

Functions of isoenzymes:

(1) Regulation of metabolic activities.

(2) Catalysis of reversible reaction.

(3) Spatial separation of the biochemical reaction.

(4) Differential feedback inhibition.

Properties of enzymes:

As enzymes are proteinaceous they show some of protein related properties. These properties help the enzymes to act more efficiently in favourable conditions. Due to these properties the enzymes accelerate the rate of reaction by lowering the energy of activation.

Some of the properties are given below:

(a) Protein nature: Enzymes are proteinaceous in nature. They are globular proteins. They need some additional organic and inorganic substances for their activity. There are some RNA enzymes.

(b) Biocatalysts: Enzymes are biocatalysts. They accelerate the rate of reaction by decreasing the energy of activation. Due to this they are required in very small amounts and remain unchanged at the end of the reaction.

(c) Specificity in action: Enzymes are very much specific about the substrates.

There are more than 700 different types of enzymes and all are specific in their action. Depending upon the substrate, enzymes show following specificity characters:

(i) **Absolute substrate specificity:** Some of the enzymes act only on one kind of substrate.

(ii) **Absolute group specificity:** Some enzymes act on several different substrates which are placed in one chemical group.

(iii) **Relative group specificity:** Some of the enzymes act on one class of compounds but also attack on another class to some extent (More than one organic group).

(iv) **Stereo chemical specificity:** Stereo chemical enzymes catalyse only those enzymatic reactions, that are involving stereochemical compounds.

e.g. D-stereo chemical enzyme act on D-stereo isomers and L-stereo chemical enzyme act on L stereo isomers.

(d) Colloidal nature: As enzymes are colloidal in nature. They provide larger surface area for the reaction to take place. As they are colloidal, they are hydrophilic in nature and form hydrosols in the free state.

(e) Thermolability: Enzymes are very sensitive to temperature. They are more active at optimum temperature and become inactive at high temperature.

(f) pH Sensitivity: Enzymes are sensitive to pH. They are more active at optimum or neutral pH and remain inactive at very high and very low pH.

(g) Reversibility: All enzyme controlled reactions are reversible. The reversibility depend upon energy requirements, availability of reactants, concentration of end products and pH.

(h) Chain reactions: Biochemical reactions are not isolated. Number of reactions takes place in sequence. A team of enzymes work one after the other to accomplish such multistep reactions.

(i) Enzyme inhibitors (protein poisons): Enzymes are inactivated or denatured by all those substances and forces which destroy protein structure. e.g. heavy metals, high energy radiations.

1.4 MECHANISM OF ENZYME ACTION: LOCK AND KEY HYPOTHESIS AND INDUCED FIT HYPOTHESIS

Mechanism of enzyme action:

As enzymes are proteins they also show some specific requirements during their action. The basic requirements of enzymes action are selective substrate, proper medium, favourable temperature and pH. The specificity and recovery in the form of product and enzyme at the end of a biochemical reaction are some of the important aspects of enzymatic reactions. In simple words the mechanism of enzyme action can be given below.

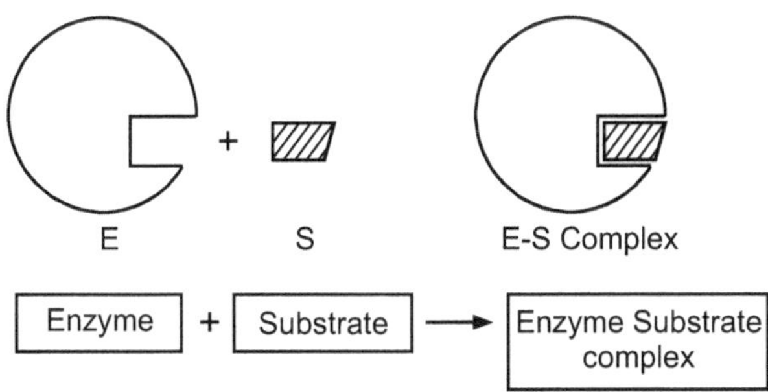

Fig. 1.9: Formation of Enzyme-substrate complex

In the beginning of the reaction a specific enzyme select a specific substrate and develop binding with it. The binding is carried out through the 'active site'. This binding is responsible to develop enzyme-substrate complex. This complex is more stable and can remain as it is for a stipulated time.

During this binding period the interaction between enzyme and substrate takes place. During these interactions the Van-der-Waals forces, hydrogen bonds and other ions are taking part in it. Due to the contribution of all these factors there is a sudden decrease in energy of activation. Hence the reaction become much faster. This condition create changes in the substrate and produces the final product.

At the end of the reaction the enzyme releases from the end product and becomes free. Hence at the end of the reaction the enzyme remains unchanged qualitatively and quantitatively.

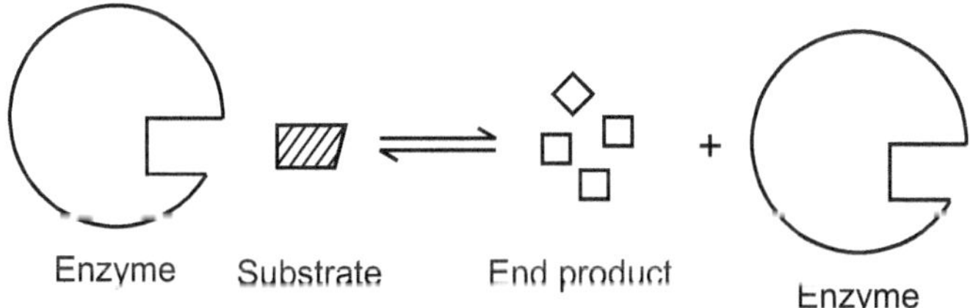

Fig. 1.10: End of the enzymatic reaction

The released enzyme again bind with new substrate. This process continues until all the substrate is converted into the product.

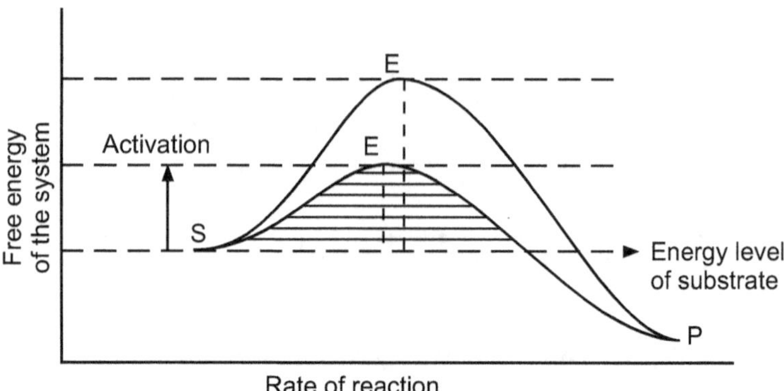

Fig. 1.11: Change in energy activation during enzymatic reaction

In short, the mechanism of enzyme action can be summarised in the following schematic flow diagrams.

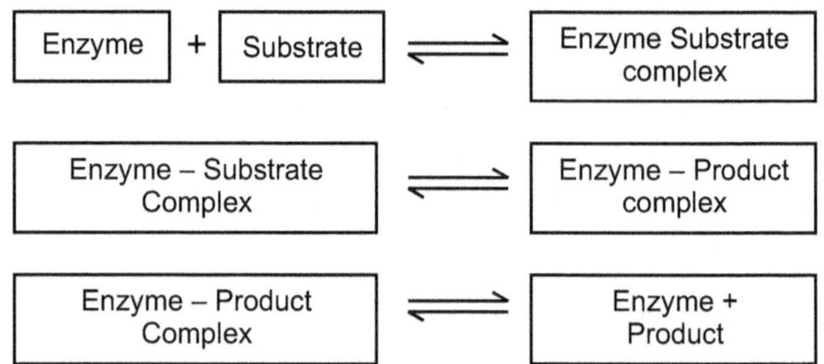

Fig. 1.12

Theories to explain mechanism of enzyme action: There are many theories put forth to explain the mechanism of enzyme action. Two theories are explained below.

(a) Lock and key theory and

(b) Induced fit theory.

(a) Lock and key theory (Hypothesis):

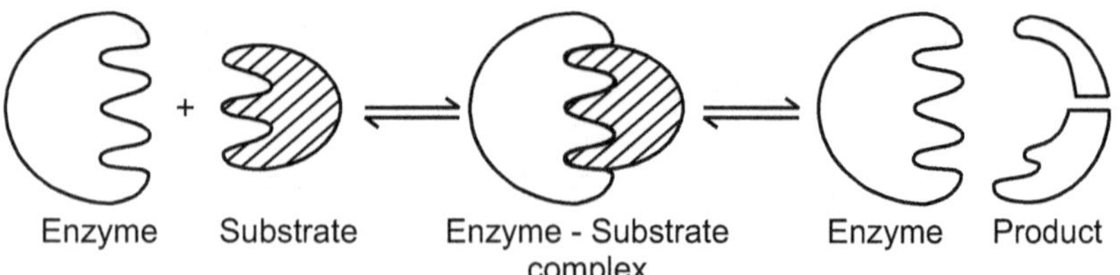

Fig. 1.13

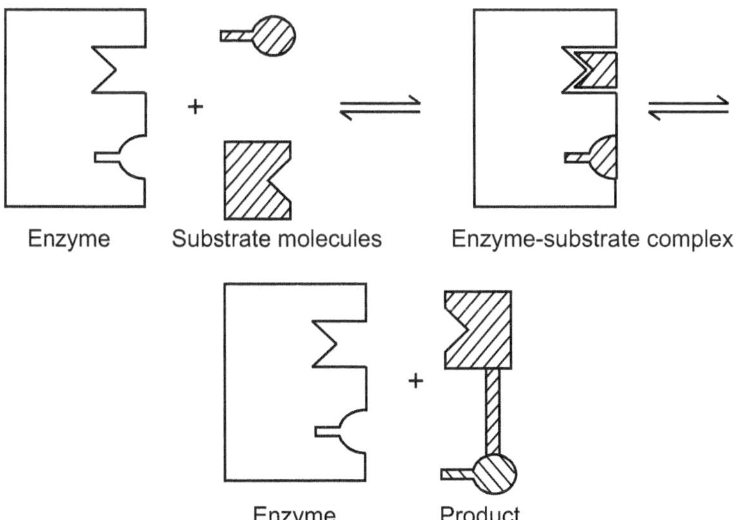

Fig. 1.14: Diagramatic representation of Lock and key hypothesis (Emil Fischer, 1890)

This hypothesis was proposed by scientist Emil Fischer in 1890. According to this hypothesis the shape of the active site is rigid or fix and it has special geometrical shapes. Similarly the shape of the active site is complementary with the shape of the substrate. The surface configuration of the active site of enzyme allow the particular substrate molecule to fit it. This action is very much similar to a common 'Lock and Key' concept. Just like specific lock will be opened by a specific key. In addition to the complementary shape the active site also has some special groups like NH_2, COOH and SH to develop a contact with the substrate molecule.

After coming in contact a particular enzyme and a particular substrate an enzyme-substrate complex will develop. The substrate undergo some conformational changes and finally product will be produced. Finally the enzyme become free to bind with another substrate molecule.

This theory helps to explain how a small quantity of enzyme can act upon a large amount of substrate. It also explains how a substrate having a structure similar to the substrate can work as competitive inhibitor.

Drawback: This hypothesis has some drawbacks like not explaining all the aspects of enzyme action, retaining the original structure or shape of active site and binding with substrates having low molecular weight.

(b) Induced Fit Theory (hypothesis):

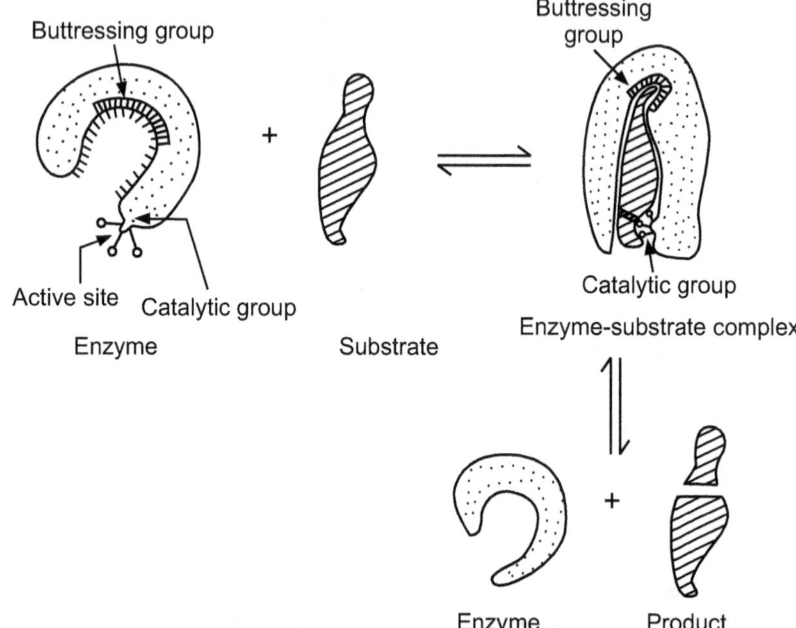

Fig. 1.15: Diagrammatic representation of Induced fit hypothesis (Koshland, 1959)

This hypothesis is a modification of lock and key hypothesis. This hypothesis was proposed by scientist Koshland in 1959. According to this hypothesis the active site of an enzyme is not fix or rigid. It is flexible and can be changed complementing to the shape of substrate.

This hypothesis also describes that the active site of an enzyme contains two groups (i) buttressing or supporting group, (ii) catalytic or interacting group. The buttressing group gives the support to the substrate to hold active site of enzyme. The catalytic group is responsible for weakening the bonds of reactions by electrophilic and nucleophilic forces.

In the beginning both the groups are apart from each other. As soon as the substrate comes in contact with the buttressing group, the active site of the enzyme undergoes some conformational changes, so as to bring the catalytic group at the opposite side of the substrate to break the bonds.

The catalytic group of the active site of enzyme chemically interact with the substrate. At the end of the interaction the substrate is converted into product. The newly formed product has totally new shape and it is not fitting with the shape of the active site. This cause the buttressing group to revert (undo) to its original position. This cause the release of the product. The freed enzyme can again act on new substrate by the same way.

Merits: This hypothesis has been proved by many advance techniques like optical rotational measurement and x-ray diffraction analysis. This hypothesis is widely accepted.

1.5　FACTOR AFFECTING ENZYME ACTIVITY- TEMPARATURE AND pH

As the enzymes are proteinaceous in nature, their activity has been affected or governed by variety of factors playing important role in biochemical reactions. These are substrate concentration, enzyme concentration, temperature, pH, enzyme inhibitors and accumulation of end product etc. Two major factors affecting the enzyme activity are temperature and pH which are explained in detail below:

A.　Temperature:

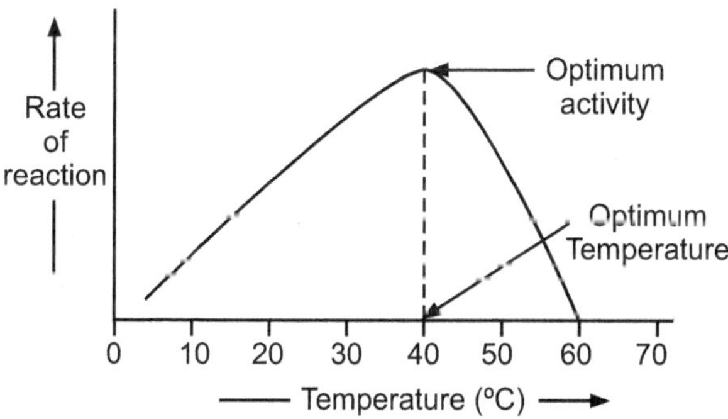

Fig. 1.16: Effect of temperature on enzyme activity

Temperature is one of the important factor affecting the enzyme activity. As the enzymes are proteinaceous in nature very high (beyond 60°C) and very low temperature (below 0°C) are adversely affect the enzyme activity. Enzyme activity increases when the temperature increases. At optimum levels (20°C to 40°C) the enzyme show maximum activity. If the temperature increase beyond optimum (45°C) the enzyme activity starts to decline and at a very high temperature it stops. At very high temperature the enzymes gets denatured and loose their catalytic property. The globular 3D enzyme structure is lost and the amino acids get separated. This is irreversible process. Temperature sensitivity depends upon the water content of the medium. The enzymes present in seeds or spores and thermophilic bacteria can remain active at high temperatures. The enzymes like ribonuclease, lecithins are tolerant to temperature upto 70°C.

B. pH:

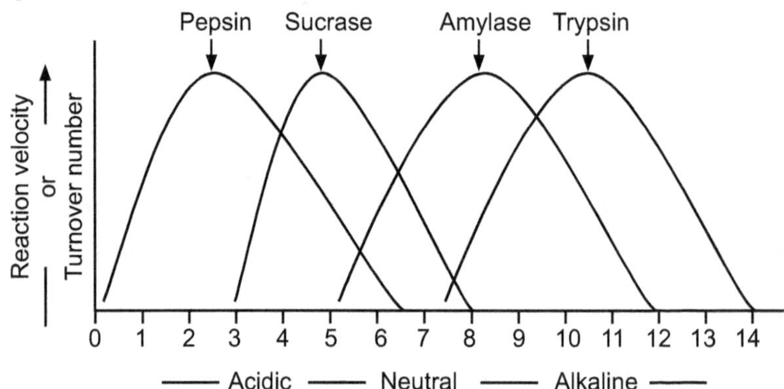

Fig. 1.17: Relation between pH and activity of enzymes

pH is the another factor affecting enzyme activity. Enzymes show optimum activity at a limited range of pH. This is because the active site of the enzyme should be in a proper ionised state to be active. The ionization of active site changes as pH changes. All the enzymes are pH specific. The enzyme show a bell shaped curve over the range. This indicates that the rate of reaction increases upto optimum pH value, beyond this value the activity declines and finally the reaction stops due to lack of enzyme activity. Specificity of pH for enzyme activity is useful in regulating the enzymes. e.g. pepsin (2 pH), sucrase (4.5 pH), salivary amylase (6.8 pH), trypsin (8.5 pH).

1.6 ENZYME INHIBITION

Reduction or stoppage of enzyme activity due to the presence of adverse conditions or chemicals is called as 'enzyme inhibition'. There are several enzyme inhibitors. Enzyme inhibition can be classified into two groups. (a) Reversible and Irreversible and (b) Competitive and Non-competitive.

The four common types of enzyme inhibition methods are described below:

(a) Protein denaturation.

(b) Competitive inhibition.

(c) Non-competitive inhibition.

(d) Allosteric modulation or feedback inhibition.

(a) Protein denaturation:

As the enzymes are proteins, their activity is affected by various factors. The enzymes remain active when it is in 3 Dimensional form. This 3D structure is disturbed by several factors like heat, high energy radiations and salts of heavy metals.

(b) Competitive inhibition:

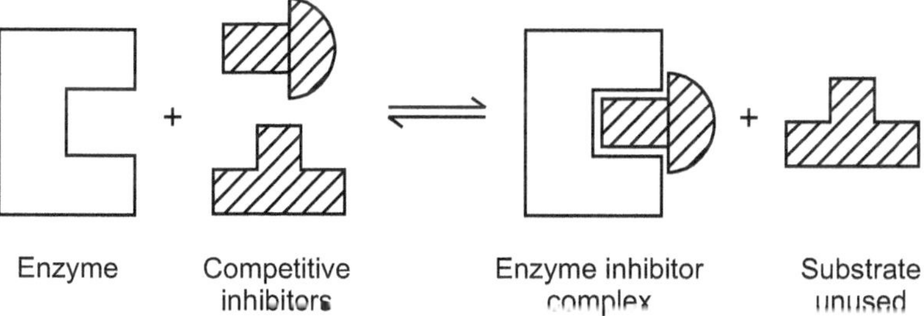

Fig. 1.18: Competitive inhibition of enzyme action

In this type of enzyme inhibition certain chemicals compete with the substrate to bind with active site of the enzyme. The inhibitor chemical is also known as substrate analog or competitive inhibitor. The chemical inhibitor has a similar shape and size like that of the substrate molecule. But when the inhibitor attaches to the enzyme it is not converted into product, due to this the enzyme cannot participate in catalytic change of the substrate. As the substrate is not converted into product chemical inhibits the catalytic process. This type of inhibition is reversible.

(c) Non-competitive inhibition:

This is an irreversible type of inhibition. In this type the chemical or substrate inhibiting the enzyme activity has no structural similarity with the substrate. The irreversible non-competitive inhibitor destroys or combines irreversibly with the functional group of enzyme, that is very much essential for its catalytic function. e.g. Cyanide inhibition.

(d) Allosteric modulation or feedback inhibition:

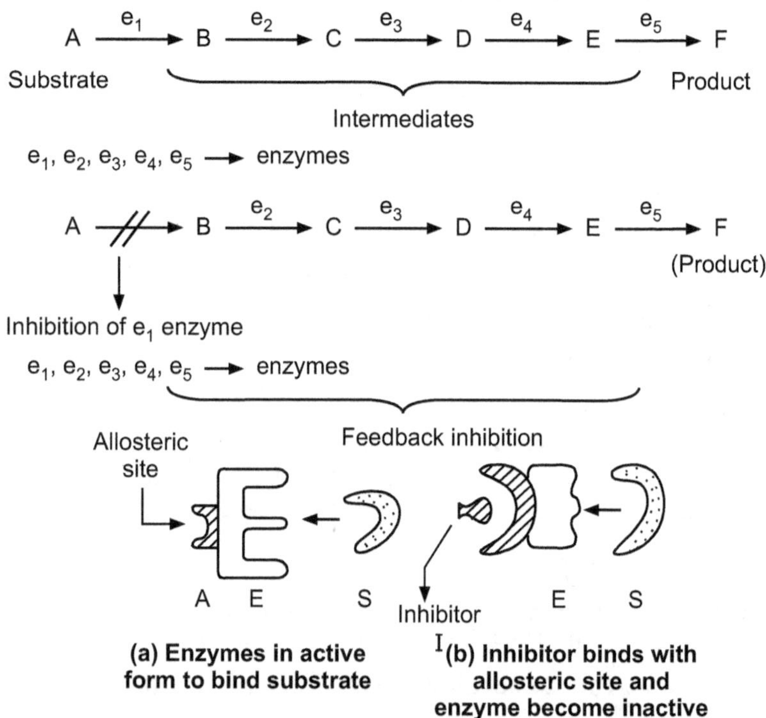

(a) Enzymes in active form to bind substrate　　**(b) Inhibitor binds with allosteric site and enzyme become inactive**

Fig. 1.19: Allosteric Feedback inhibition of enzyme activity

Enzymatic reactions are continuously taking place. Due to which there is continuous product development. This causes more accumulation of product in the medium (cell). This may cause damage to the cell. There is a need to stop the enzymatic activity automatically. The end product itself is responsible to stop the reaction. There is a systematic mechanism to stop the enzyme activity. This mechanism is called as Allosteric modulation or Allosteric feedback inhibition. (Allosteric Greek meaning- allo – other, stereos – space or site). During this process an additional regulator or inhibitor molecule is automatically synthesized. This regulator molecule has totally different shape than the active site. This inhibitor molecule attaches the enzyme molecule on the opposite side of the active site. Due to this inhibitor attachment the enzyme surface gets stretched. Due to stretching, the shape of the active site get changed conformationally. Such an enzyme with two sub-units (one active and other inactive) is called Allosteric enzyme. Due to the inhibitor molecule the enzyme cannot bind with the substrate molecule. Hence Enzyme-Substrate complex is not developed. Ultimately no end product is synthesized. If this allosteric mechanism takes place in the early steps of sequential reactions (e_1) the remaining reactions will be automatically stopped and the end product will not be synthesized. Allosteric inhibition is particularly important in end-product inhibition.

IMPORTANCE

(1)   Play regulatory role in enzyme action.
(2)   Used in study of metabolic pathway.
(3)   Used in control of some pathogens e.g. Sulfa drugs.

QUESTIONS

I.   Broad Questions:

1. What are enzymes? Give the classification of enzymes.
2. What are enzymes? Give the properties of enzymes.
3. Describe mechanism of enzyme action.
4. Explain Lock and key theory and Induced Fit theory.
5. Explain the effect of temperature and pH on enzyme activity.
6. Explain Allosteric feedback inhibition of enzymes.

II. Short Notes:

1. Nomenclature of enzymes.
2. Primary and Secondary structure of enzymes.
3. Tertiary structure of enzymes.
4. Conjugated enzymes.
5. Co-enzymes
6. Co-factors
7. Isoenzymes

III. Multiple choice Questions:

1. Chemically enzymes are ....

   (a) proteins   (b) lipids

   (c) carbohydrates   (d) fats

2. Enzymes are ....

   (a) special fats   (b) special proteins

   (c) special lipids   (d) special carbohydrates

3. Holoenzymes means ....

   (a) inhibitor

   (b) non-protein part

   (c) protein part and non protein part

   (d) accelerator

4. Isoenzymes are ....

   (a) multiple forms of enzymes

   (b) multiple forms of lipids

   (c) multiple forms of fats

   (d) multiple forms of carbohydrates

5. The energy of activation .... by addition of an enzyme in biochemical reaction.

   (a) increases   (b) decreases

   (c) neutral   (d) suddenly increased

6. At very high temperature the enzyme get ....

   (a) disappeared   (b) reappeared

   (c) denatured   (d) active

7. Non-protein part of enzyme is called as ....
    (a) cis   (b) trans
    (c) aldo   (d) prosthetic group
8. Lock and key theory was proposed by ....
    (a) E. Fischer   (b) Koshland
    (c) Kuhne   (d) Moller
9. Induced fit theory was proposed by ....
    (a) Fischer   (b) Kuhne
    (c) Moller   (d) Koshland
10. IUB means ....
    (a) International Union of Broadcasting
    (b) International Union of Biochemistry
    (c) International Union of Botany
    (d) International Union of Business
11. There are .... major classes of enzymes.
    (a) six   (b) seven
    (c) four   (d) five
12. There are .... ends to each 3D enzyme structure.
    (a) four   (b) five
    (c) ten   (d) two
13. The enzyme code consist .... digits.
    (a) eight   (b) four
    (c) three   (d) six
14. The tertiary enzyme structure is also called ....
    (a) 3D   (b) 4D
    (c) 5D   (d) 6D
15. The quaternary structure of enzymes are also called as
    ....
    (a) prosthetic group   (b) holoenzyme
    (c) apoenzyme   (d) isoenzymes
16. The metal ions attached to enzymes are called ....
    (a) co-enzymes   (b) apoenzymes
    (c) co-factors   (d) holoenzymes

17. Multiple forms of enzymes are called as ....

    (a) isoenzymes   (b) co-factors

    (c) co-enzymes   (d) metal ions

18. Feedback inhibition of enzyme is also called as ....

    (a) bisteric   (b) allosteric

    (c) tristeric   (d) monosteric

Answer key:

(1) – a, (2) – b, (3) – c, (4) – a, (5) – b, (6) – c, (7) – d, (8) – a, (9) – d, (10) – b, (11) – a, (12) – d, (13) – b, (14) – a, (15) – b, (16) – c, (17) – a, (18) – b.

NITROGEN METABOLISM

2.1 INTRODUCTION

Nitrogen is a naturally occurring element, which is essential for growth and reproduction in both plants and animals. It is a component of amino acids that make up proteins and nucleiac acids, which constitute hereditary material. It is also present in many organic and inorganic compounds. It comprise 78% of the earths atmosphere. It is interesting to note that nitrogen is present in the atmosphere in such a large amount, but out of which very less amount is available to the plants and other organisms. For this it is necessary to convert inorganic nitrogen into soluble organic forms. This process takes place inside the plants. Plant cells are the only entity having ability to convert the inorganic nitrogen into organic nitrogen.

Nitrogen plays an important role as the backbone of many other organic compounds like chlorophylls, cytochromes, alkaloids and many vitamins. In short nitrogen plays fundamental role in metabolism, growth, reproduction, heredity, cell division and as a component of energy rich molecules like ATP, GTP etc.

Sources of nitrogen to the plants :

(1) Atmospheric Nitrogen (Molecular Nitrogen N_2):

The amount of atmospheric nitrogen is 78%. It is present in molecular form (N_2). It is not used by the plants directly. Only some bacteria, blue green algae (BGA) and some legumes can fix atmospheric N_2.

(2) Nitrates, Nitrites and Ammonia:

The nitrogen is available in various forms. Out of these forms nitrate is water soluble and easily absorbed by the plants. The artificial fertilizers are also prepared in this form.

(3) Amino acids in the soil:

Nitrogen can also be present in the form of amino acids in the soil. The micro organisms present in the soil use this form of nitrogen.

(4) Organic nitrogen compounds in the insects:

Many times some plants cannot fulfil their nitrogen requirement through soil. They show a separate adaptations for this purpose. These are called as 'Insectivorous plants'. These plants trap the pollinating insects with the help of special digestive glands. These glands consist digestive enzymes. These plants secrete the nitrogen from the body of the insects, and fulfil their nitrogen requirement. In short, N_2 metabolism can be defined as "It is a chain of biochemical and physical processes taking place in collaboration with physical and biological factors. During the process various complicated organic forms of N_2 are produced by using inorganic forms and physical activities. These complex forms are degraded into their simple nitrogen compounds".

2.2 BIOLOGICAL NITROGEN FIXATION

Biological Nitrogen Fixation can be defined as "It is the conversion of molecular N_2 into ammonia by microorganisms is known as biological nitrogen fixation". About 90% of the total nitrogen has been contributed by this method. This nitrogen circulation is carried out under some basic requirements like temperature and some enzymes like nitrogenase. The N_2 fixing microorganisms are also called diazotrophs.

$$N_2 + 3\,H_2 \xrightarrow{\text{Nitrogenase}} 2NH_3$$

Biological N_2 fixation can be categorized in two ways.

(a) Asymbiotic N_2 Fixation

(b) Symbiotic N_2 Fixation

(a) Asymbiotic N_2 Fixation:

Asymbiotic N_2 Fixation can be defined as "the N_2 fixation carried out by free living microorganisms directly without the help of any other organisms." All the N_2 fixing organisms are prokaryotes (bacteria). They live independently of other organisms.

Many scientist like **Shneider et.al** (1960), **Carnahan et.al.** (1960) tried to explain asymbiotic N_2 fixation by using N_2 fixing bacteria and Blue Green Algae (BGA) using isotope of nitrogen N^{15}. Microorganisms which pass independent life and fix atmospheric N_2 are known as free living diazotrophs. There are three groups of such organisms. These are:

(i) Free living bacteria

(ii) Blue Green Algae (BGA)

(iii) Fungi

(i) Free living bacteria: Based on the mode of nutrition (carbon, nitrogen, oxygen and requirement of reducing groups) bacteria are divided into (a) aerobic (obligate), (b) facultative and (c) anerobic. Free living bacteria means that they don't need to create symbiotic relationships with plants to survive and replicate. The nitrogen fixation by free living bacteria contribute a lot to the nitrogen cycle. Following is the list of free living N_2 fixing bacteria occurring in the nature.

Table 2.1: Examples of Free Living N2 Bacteria

Obligate	Facultatively anaerobic	Anaerobic
Azatobacter Chroococcum	*Klebsiella pneumonia*	*Clostridium pasteurianum*
A. Vinelandii	*Bacillus polymyxa*	*Chlorobium limicola*
Azomonas agilis	*Pseudomonas sp.*	*Chromatium okenii*
Achromobacter		*Desulfovibrio desulfuricans*
Arthobacter globiformis		*Methanobacterium formicicum*
Azospirillum lipoferum		*Methanosarcina barkeri*
		Rhodomicrobium vammielii
		Rhodospirillum rubrum
		Rhodobacter capsulants
		Heliobacterium chlorum
		Methylomonas communis

(ii) Blue Green Algae (BGA): BGA is a major group of microorganisms who fix the atmospheric N_2 directly without the help of the other. There are various forms of BGA like unicellular, colonial, fillamentous etc. The filamentous forms are further sub-grouped as, (a) Filamentous with heterocyst, (b) Filamentous without heterocyst.

(a) Filamentous with heterocyst: e.g. *Nostoc, Anabena, Cylindrospermum, Tolypothrix.*

(b) Filamentous without heterocyst: e.g. *Oscillatoria, Trichodesmium, Lyngbya.*

During asymbiotic N_2 fixation enzyme nitrogenase plays important role. Ferredoxin plays the role of reducing agent. The ATP is used and ammonia is synthesized from N_2 in presence of enzyme nitrogenase. ATP is given by pyruvate along with H^+ and e^- via $NADH_2$ and Ferredoxin. To convert one N_2 molecule into two NH_3 molecules, require 1.5 ATP, 6 H^+ (ions) and 6 e^- (electrons).

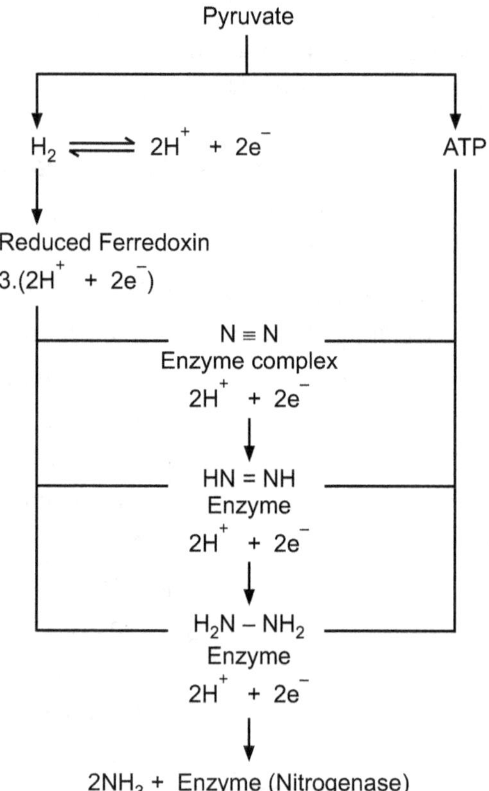

Fig. 2.1: Mechanism of N_2 Fixation in Blue Green Algae

(iii) Fungi: There are very few fungal species which have the capacity of fixing Nitrogen. e.g. *Saccharomyces, Mycoderma vini, Aspergillus niger, Penicillium glaucum* and *Botrytis cinerea.*

(b) Symbiotic Nitrogen Fixation:

Symbiosis is a type of an association in which both the components are beneficial to each other. The reaction between N_2 Fixing bacteria and plants is the best example of this. These N_2 Fixing bacteria develop a strong symbiotic association with the roots of leguminous plants and develop a globular dark red coloured clump or knot like structure called as the 'nodules' In this association the plants provide carbohydrates and support to the N_2 fixing bacteria, in return N_2 fixing bacteria gives fixed N_2 to the host plants. These N_2 fixing bacteria infect the soil for a longer period, which gives benefit to the next crop also. This cause the soil more fertile. Hence the crop rotation with the leguminous plants is very much beneficial to maintain the nitrogen content of the soil.

The symbiotic N_2 fixation can be categorized as follows.

(i) Root nodule forming leguminous plants- e.g. *Rhizobium, Frankia.*

(ii) Root nodule forming non-leguminous plants – e.g. *Beijerinckia, Azatobacter.*

(iii) Symbiotic Blue Green Algae – e.g. *Nostoc, Anabaena, Calothrix, Haplosiphon.*

(i) Root nodule forming leguminous plants: The leguminous plants and N_2 Fixing bacteria develop an unique association. The process is complex and proves successful and beneficial if the requisite environment is provided. The process of nodulation in leguminous plants can be explained in two steps:

(a) Nodule Formation (Nodulation)

(b) Mechanism of N_2 Fixation by - root nodules.

(a) Nodulation (nodule formation):

The root nodule formation (nodulation) process in leguminous plants is an interesting process. This process is carried out in a sequence, like-infestation of soil by bacteria, recognition of particular bacterial strain by lecithin protein, infection by the bacteria in root hair and finally development of mature root nodule.

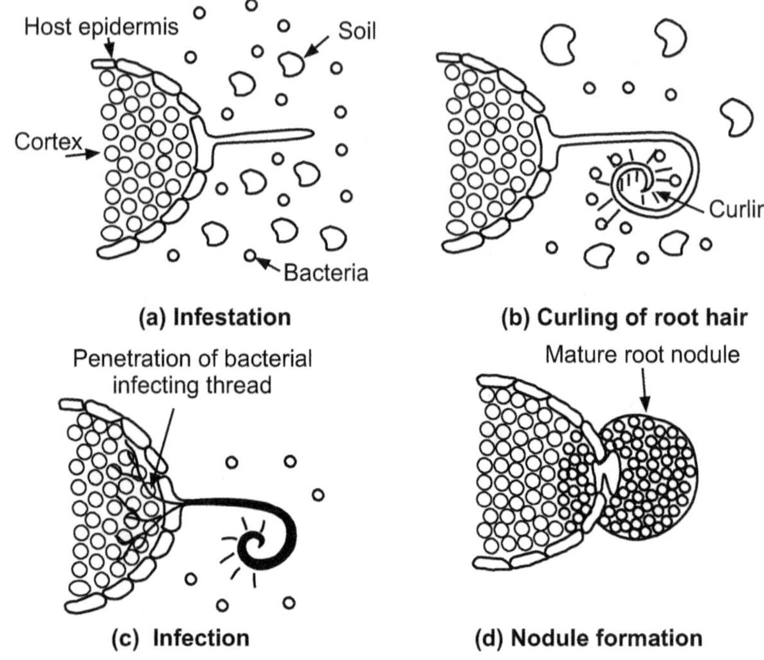

Fig. 2.2: Nodulation

Various types of bacteria are present in the soil. The number of naturally occurring N_2 fixing bacterial cells in the soil is variable in number. It may be less in newly formed soils. The number of N_2 fixing bacterial cells in the soil can be increased by artificially infesting (inoculating) the soil or legume seeds with the N_2 fixing bacteria like *Rhizobium* or *Azatobacter*. According to some views the detection of N_2 fixing bacteria from the soil by the legume roots is a genetically controlled process. Some genes may be responsible for this.

The root cells detect the requisite N_2 fixing bacterial strains from the soil. (Specific leguminous plants/roots can pick up specific bacterial strains.) To attract the bacteria the genes present in the root hairs, secrete specific proteins like 'lecithin'. The lecithin detect the glycoproteins, cytokinins or polymyxin − B from the bacterial cells. They interact each other and develop 'Rhizobium-legume' association. Calcium is essential for this process.

The bacterial cells infect the thin walled epidermal cells of root hairs made up of cellulose. The roots release the chemical 'tryptophan' to attract the bacteria. In return bacteria secretes more auxin like IAA or cytokinins, which cause the curling or bending of the root hairs. This favours large amount of bacteria to enter inside the roots. The rhizobia enter the root cortex in the form of tiny bodies called 'bacteroids'. The bacteroid induce multiplication of cortical cells to form the tetraploid (4n) cells. The repeated cell division cause the formation of round shaped dark purple or red coloured 'root nodules'. The dark red colour is due to the presence of pigment 'leghaemoglobin', which combines with O_2 and gets oxidised into brown form with Fe^{++} ions.

(b) Mechanism of N_2 Fixation by root nodules:

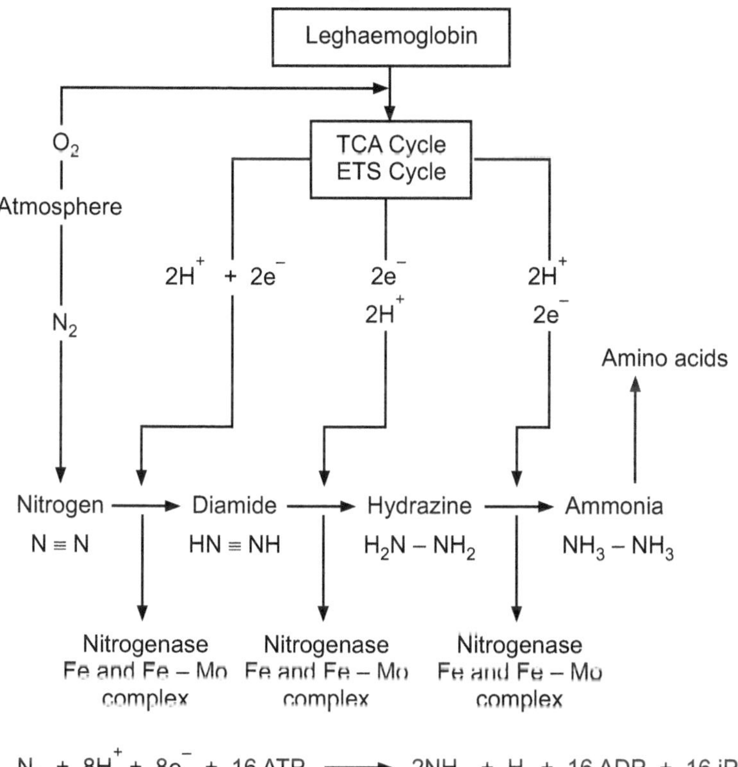

$$N_2 + 8H^+ + 8e^- + 16\,ATP \longrightarrow 2NH_3 + H_2 + 16\,ADP + 16\,iP$$

Fig. 2.3: Mechanism of N_2 Fixation in Root Nodules

During this process the nitrogen fixed by N_2 fixing bacterias or root nodules is finally converted into two molecules of ammonia (NH_3) by forming diamide hydrazine as the intermediates. The dark red pigment 'leghaemoglobin' have the capacity to fix nitrogen. Leghaemoglobin regulates the O_2 level, which is required for the activity of enzyme 'Nitrogenase'.

By combining with O_2 leghaemoglobin produces oxyhaemoglobin. This cause the decrease in O_2 Level. The N_2 bound to the surface of enzyme nitrogenase and is reduced in step wise reactions.

The reduction of N_2 to ammonia in root nodules require ATP and reduced substrate capable of donating hydrogen atoms into N_2. ATP generation takes place via. respiration of bacteria. The substrate required by the enzyme activity is Glucose – 6-phosphate. NADP with ferredoxin act as electron carrier for the process. The ATP interacts with protein component of nitrogenase, which cause the reductional changes to reduce N_2 to NH_3.

Four molecules of ATP are hydrolysed for each pair of electrons transferred to nitrogen. Reduction of each molecule of N_2, into two molecules of ammonia (NH_3) require 12 ATP Molecules as 6 electrons are required per molecule of N_2 reduced. The synthesized NH_3 is later on immediately converted into Amino acids. Because excess concentration of NH_3 is harmful to the cell. E.g. *Rhizobium, Frankia*.

(ii) Root nodule forming non - leguminous plants: There are some non-legume which Fix N_2 symbiotically. These are - *Casuarina, Coenothus, Cycus, Podocarpus, Alnus* and *Myrica*.

(iii) Symbiotic Blue Green Algae (BGA): Some members of BGA show association with other plants. The BGA like *Nostoc, Anabaena Haplosiphon, Calothrix* shows the association with plants like *Cycas, Zamia, Macrozamią, Anthoceros, Sphagnum, Azolla* and fix atmospheric N_2 directly.

2.3 MECHANISM OF NITROGEN FIXATION

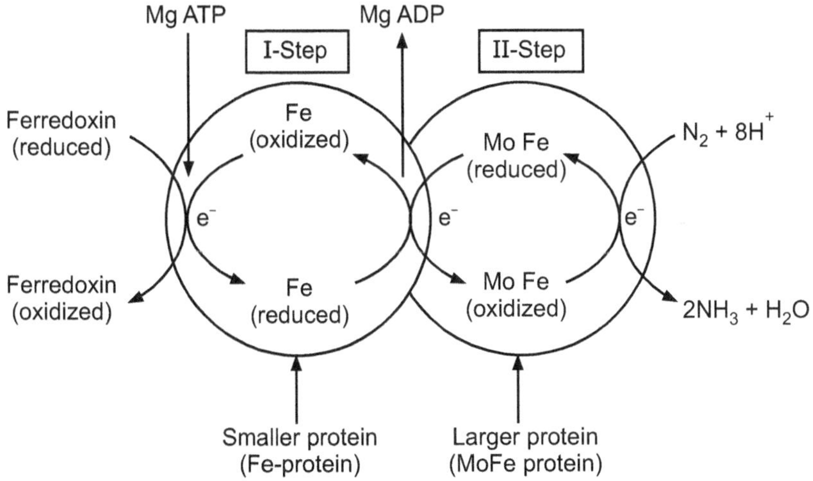

Fig. 2.4: Structure of Enzyme Nitrogenase

The conversion of N_2 to ammonia is a complex process and the reaction is catalysed by enzyme nitrogenase. Nitrogenase show two protein components. One unit contain iron (Fe) and molybdenum (Mo) known as MoFe complex and other containing only Fe protein and called Fe protein. The Mo-Fe protein consist 30 to 32 iron atoms and 1 to 2 atoms of Mo. The enzyme nitrogenase is very sensitive to O_2 and at the same time O_2 is the inhibitor of N_2 fixation. The pigment leghaemoglobin plays important role to protect activity of enzyme nitrogenase. The leghaemoglobin regulates the O_2 level to prepare the favourable environment or medium for nitrogenase activity.

2.4 NITRATE REDUCTION (NITRATE ASSIMILATION)

Nitrate (NO_3^-) is an important form of N_2 produced during various photo chemical reactions taking place during N_2 Fixation. It is available from the soil. This is present in the oxidized (NO_3^-) form which is later on finally reduced to Ammonia (NH_4^+). These are water soluble and absorbed by the roots. In root cell cytoplasm they are converted into ammonia (NH_3). This process is called as Nitrate reduction or Nitrate assimilation.

This reduction process carried out in two steps and the enzymes catalyzing these reactions are nitrate 'reductase and nitrite reductase. Chemically these are metaloflavoproteins. At each step two electrons are added and ultimately NO_3 is produced in which nitrogen has 3 negative charges. The electrons are supplied by reduced co enzyme - I (NADH) and co-enzyme-II (NADPH). The two steps are explained as below.

First step - Reduction of Nitrate to Nitrite (In cytoplasm):

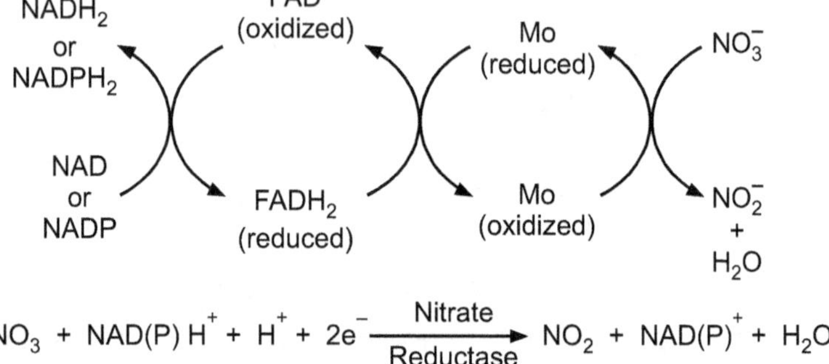

$$NO_3 + NAD(P)\,H^+ + H^+ + 2e^- \xrightarrow[\text{Reductase}]{\text{Nitrate}} NO_2 + NAD(P)^+ + H_2O$$

Fig. 2.5: Reduction of nitrate to nitrite

Reduction of nitrate to nitrite takes place in the presence of the enzyme nitrate reductase. It require co-enzyme - I (NADH) or co-enzyme-II (NADPH). This enzyme is Molybdoflavoprotein with an operative sulfhydryl group. This enzyme contain FAD as its prosthetic group with the help of this it interact with Mo (molybdenum). Electrons are transferred from reduced co-enzyme to FAD which become reduced $FADH_2$. From reduced $FADH_2$, the electrons are finally transferred to NO_3 through molybdenum so that NO_2 and H_2O are formed. Nitrate reduction is takes place in green leaves and roots.

Second step - Reduction of Nitrite to Ammonia (In root and leaves)

$$3\,NADH + NO_2 + 4H^+ \xrightarrow[\text{reductase}]{\text{nitrite}} 3NAD + NH_3 + 2H_2O$$

The product of the first step NO_2^- is further reduced to NH_3 in the second step. This reaction takes place either in root or the leaves. The reaction is catalysed by the enzyme 'Nitrite reductase' (NiR). This enzyme requires reduced co-enzyme I and II. (NADH and NADPH).

These co-enzymes supply the hydrogen atoms. The Manganese (Mn) act as the prosthetic group for this enzyme. Presence of enzyme Nitrite reductase is necessary for the assimilation of nitrogen in plants. In absence of this enzyme nitrogen will not be assimilated and the plants can show nitrogen deficiency symptoms.

2.5 AMMONIA ASSIMILATION

The end product of the nitrate reduction is ammonia (NH_3). The ammonia has been synthesized as a result of nitrogen fixation, nitrate reduction and ammonification in plants. The ammonia can also be available from the soil. There are many soils particularly detritus, where there is a large activity of decomposition and microorganisms. Such soils are observed in mangroves. The accumulation of excess ammonia is harmful to the cell. This harmful ammonia can be converted into harmless substances like amino acids and water. The excess amount of ammonia can also affect the activity of the enzyme nitrogenase, energy processes and ETS. To avoid these toxic effects of ammonia it is essential to convert the ammonia into amino acids. This process is known as 'Ammonia assimilation'. The living organisms can assimilate ammonia in organic compounds by many reactions like glutamate dehydrogenase reaction, glutamine synthetase reaction, carbamyl phosphate synthetase reactions.

The ammonia in organisms is mostly utilized for the synthesis of glutamine which serves as the major nitrogen donor in biosynthetic reactions. In addition carbamyl phosphate donates nitrogen for the biosynthesis of arginine and pyrimidines etc.

Assimilation of ammonia or amino acid synthesis is carried out by many reactions. The two major reactions are (a) Reductive amination, (b) Transamination reaction. In addition to these, it is also carried out by other ways like synthesis of carbamyl phosphate, amide synthesis and GS-GOGAT cycle.

(a) Reductive amination:

$$\text{α-Ketoglutaric acid} + NADPH_2 + NH_3 \xrightarrow{\text{Glutamic dehydrogenase}} \text{Glutamic acid} + NADP + H_2O$$

Inorganic nitrogen in the form of NH_3 produced from various activities reacts with α-ketoglutaric acid in presence of enzyme glutamic dehydrogenase and reduced co-enzyme $NADPH_2$ to form an amino acid called glutamic acid.

As this process of conversion of NH_3 into amino acid is accompanied by amination and reduction at the keto group of the organic acid, it is called as 'reductive amination'. Further the glutamic acid is converted to glutamine and the reaction is catalysed by enzyme 'Glutamine Synthetase' (GS).

$$\text{Glutamic acid} + NH_4^+ + ATP \underset{}{\overset{GS}{\rightleftharpoons}} \text{Glutamine} + ADP + iP$$

The glutamine is converted to glutamate and the reaction is catalysed by 'glutamate synthetase' (GOGAT). The glutamate act as the substrate for transamination reaction.

$$\text{Glutamine} + 2-\text{oxyglutarate} + NADH + H^+ \xrightarrow{\text{NADH-GOGAT}} 2-\text{Glutamate} + NAD^+$$

$$\text{Glutamine} + 2-\text{oxyglutarate} + Fd\,(red) \xrightarrow{\text{Fd-GOGAT}} 2-\text{Glutamate} + Fd\,(oxide)$$

In above reactions the glutamine is converted back to glutamate by the transfer of amino group. The reaction is catalysed by GOGAT.

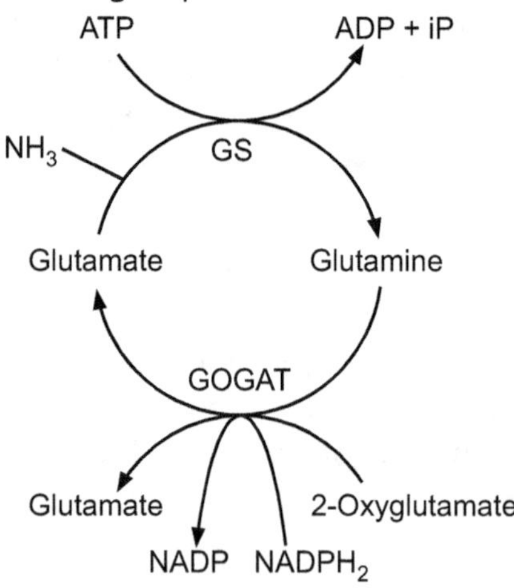

Fig. 2.6: GS – GOGAT Cyclic pattern

In ammonia assimilation the NH_4^+ is converted into glutamine by enzyme glutamate synthetase (GOGAT). This cycle shows two enzymes GS and GOGAT which carry out the reactions in cyclic manner. In this process the organic acids are reduced to amino acids by directly utilizing ammonia e.g.

(1) α-ketoglutaric acid + $NH_3 \longrightarrow$ Glutamic acid

(2) Oxaloacetic acid + $NH_3 \longrightarrow$ Aspartic acid

(3) Fumaric acid + $NH_3 \longrightarrow$ Aspartic acid

(4) Pyruvic acid + $NH_3 \longrightarrow$ Alanine

(b) Transamination:

The transfer of amino group from one amino acid to a keto acid is called 'transamination'. These reactions are catalysed by a group of enzymes called 'aminotransferases'. As a donor of amino group glutamate can produce seventeen (17) different types of amino acids.

$$\text{Glutamate + OAA} \xrightarrow{\text{transaminase}} \alpha\text{-ketoglutaric acid + Aspartic acid}$$

$$\text{Glutamate + pyruvic acid} \xrightarrow{\text{transaminase}} \alpha\text{-ketoglutaric acid + Alanine}$$

$$NH_3 + CO_2 + ATP \longrightarrow \text{Carbamyl phosphate} + ADP$$

$$\text{Carbamyl phosphate} \xrightarrow{\text{series of reactions}} \text{Arginine}$$

This pathway is also considered as the third pathway. Various amino acids are synthesized by such reactions. e.g. Glutamine, Proline, Ornithine, Alanine, Asparagine, Serine, Glycine, Hydroxy prolene, Cysteine, Alginine, Threonine, Homo-serine, Methionine, Lysine, Histidine, Phenylalanine, Tyrosine, Tryptophan, Isoleucine, Leucine, Valine etc.

2.6 NIF GENES

The research work done by various scientist proved that the process of N_2 fixation in symbiotic and free living organisms is genetical and some special genes are responsible for N_2 fixation. N_2 fixation is carried out by three groups of genes, These are :

(a) Nod genes: responsible for nodule formation

(b) Hup genes: responsible for nitrogen uptake.

(c) Nif genes: responsible for nitrogen fixation.

All these three types of genes are present in a group on a single chromosome. This makes their copying and transfer mechanism simple.

NIF genes:

The Nif genes are present in the genome of both, free living and symbiotic N_2 fixing bacteria. NiF genes are the genes encoding enzymes involved in the fixation of atmospheric nitrogen into various other nitrogen forms, which are available for living organisms. The primary enzyme encoded by the Nif genes is the enzyme nitrogenase. The nitrogenase play key role in atmospheric N_2 fixation, and converting it into ammonia. NiF genes also encode various regulatory proteins involved in N_2 Fixation. The expression of Nif genes is regulated by low concentration of fixed nitrogen and O_2 concentration.

Regulation of NiF gene activity:

The NiF genes can be found on bacterial chromosomes, but in symbiotic bacteria they are often found on plasmids. In most bacteria regulation of NiF genes transcription is done by the nitrogen sensitive NifA protein. When there is a lack of fixed N_2 for use the nitrogen regulator (NtrC) increase the expression of NifA and NifA activate all other N_2 fixing genes. If there is sufficient N_2, or O_2 another protein is activated called NiFL. This NiFL inhibit NifA activity resulting in the inhibition of activity of enzyme nitrogenase. This cause the stop of N_2 fixation.

Examples: The Nif genes in most of the N_2 Fixing organisms show some comman Features but they differ in character and qualities from one organism to another. Some of the examples are explained below.

(a) *Klebsiella pneumoniae*: This is a free living anaerobic N_2 fixing bacteria. It contains total of 20 nif genes (old number is 17) located on the chromosome in 24kb distance.

In *klebsiella pneumoniae*

(i)　niFH, niFD and niFK-encode the nitrogen as subunits.

(ii) niFE, niFN, niFU, niFS, niFV, niFW, niFX, niFB and niFQ- are responsible for adding Fe and Mo into nitrogenase complex.

(iii) niFF and niFJ- encode proteins related to electron transfer in reduction process.

(iv) niFA and niFL- act as regulatory protein and regulate other niF genes.

(b) Rhodospirillum rubrum: This is a anaerobic photosynthetic bacterium. The niF genes in this bacterium do the functions as similar as above. But in addition, niF genes also act in a metabolic way. They perform ADP - ribosylation of arginine residue in nitrogenase complex. This ribosylation is essential to activate enzyme nitrogenase when reduced nitrogen is acting as electron barrier.

(c) Rhodobacter capsulatus: This is a free living anaerobic phototroph containing transcription niF gene regulatory system. It uses different NifA activator which activates Ntrc and other nif genes.

(d) Rhizobium species: This is a gram negative N_2 fixing bacteria , which develop symbiotic relation with legumes In this process the niF genes are located on plasmids called symplasmids, (sym-symbiosis) which contains the genes related to nitrogen fixation and metabolism.

Role of Nif genes in transgenic plants:

Transfer of different genes in plants for improvement is called as transgenic plants. The transfer of niF genes is important to develop transgenic plants to solve the problem of nitrogen fertilizer supply to the plants. This is beneficial for economics and environment also. For this purpose, Ti based plasmid and Cauliflower mosaic virus based promoter (CAM promoter) was used to transfer Nif genes into non-legume plants. To test the efficacy of this system, phaseolin gene from legume (Pulses) has been transferred to sunflower where it was expressed and produce phaseolin.

Protoplast isolation from root nodules and preparations of rhizobia were used to develop hybrids by protoplast fusion and organelles uptake.

Table 2.1: NiF gene products and their functions in K. pneumoniae

Genes		Function
Q	→	Mo uptake or processing
B	→	synthesis or processing of FeMoCo.
A	→	Regulatory-nif A product activates the other operons.
L	→	Regulatory
F	→	A Flavoprotein involved in electron transfer in nitrogenase
M		Processing of Fe-protein
V		Influences specificity oF FeMoCo – protein
S		Formation of 18kd.
U		Not known
X		cloning of nif DNA
N		Help in Formation of FeMoCo Cofactor and protein for niFB.
E		similar to N
Y		similar to X
K		a subunit of nitrogenase help in transcription of nifH-D and –K, B- subunit of MoFe protein.
D		codes for α-subunit of MoFe protein.
H		a subunit of nitrogenase reductase, codes for subunit Fe protein.
J		involved in electron transfer to nitrogenase.

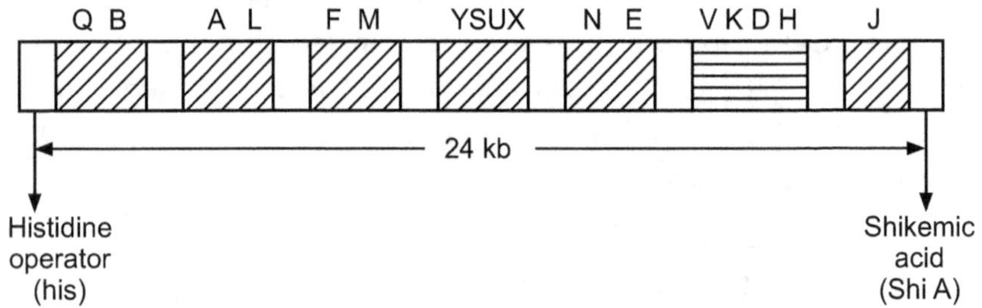

Fig. 2.7: NiF genes on Klebsiella DNA

QUESTIONS

I. Broad Questions:

1. Explain symbiotic N_2 fixation in root nodules.
2. Explain stepwise Nitrate reduction in plants.
3. Explain Reductive amination and Transamination in plants.

II. Short Notes:

1. Sources of N_2 for plants
2. Asymbiotic N_2 Fixation
3. Mechanism of N_2 Fixation in BGA
4. Root nodule formation in legumes
5. structure and function of enzyme Nitrogenase
6. Nif genes

III. Multiple choice Questions:

1. N_2 Comprises .... of the earths atmosphere.
   (a) 10%   (b) 20%
   (c) 60%   (d) 78%

2. The atmospheric nitrogen present in .... form.
   (a) N_2 (b) NO_3
   (c) NH_3 (d) NH_4

3. The N_2 fixing microorganisms are also called as ....
   (a) autotrophs   (b) diazotrophs
   (c) heterotrophs   (d) phototrophs

4. *Klebsiella pneumoniae* is an example of .... N_2 fixing bacteria.
   (a) obligate   (b) facultatively obligate
   (c) anaerobic   (d) saline

5. .... pigment is present in root nodules.
   (a) Chlorophyll   (b) Phycocyanin
   (c) Xanthophyll   (d) Leghaemoglobin

6. .... Protein is secreted in roots to attract N_2 fixing bacteria from soil.

   (a) Lecithin

   (b) Arginine

   (c) Cystene

   (d) Polymorphin

7. The enzyme nitrogenase show .... protein components.

   (a) 4

   (b) 3

   (c) 2

   (d) 1

8. The transfer of amino group from one amino acid to a keto acid is called as ....

   (a) ammonification

   (b) N_2 reduction

   (c) transamination

   (d) assimilation

9. Enzymes nitrogenase consist .... complexes.

   (a) Zn × Cu

   (b) Fe × Mo

   (c) Ni × Cd

   (d) Ca × Mg

Answer key:

(1) – d, (2) – a, (3) – b, (4) – b, (5) – d, (6) – a, (7) – c, (8) – c, (9) – b.

RESPIRATION

3.1 INTRODUCTION

Respiration is a Latin word : Re = repeatedly and spirae = to breathe.

Respiration is a vital process which occurs in all living cells of the plant but the most actively respiring regions are growing parts like floral and vegetative buds, germinating seedlings, stem and root apices. All living organisms require energy to carry out various activities. This energy is obtained through respiration which is mainly a catabolic process. Respiration is catabolic process in which complex compound (carbohydrates, proteins and fats etc.) are oxidized and to release in the form of ATP molecules. From ATP molecules, energy is supplied during various metabolic reactions. Such ATP molecules are considered as energy currency and they can rapidly supply energy to any metabolic reactions. The main aim of respiration is to produce energy currency molecules, as per the requirement of various cells and organs in the plant body.

Respiration is a process by which living cells break down complex high energy food molecules into simple low energy molecules i.e. CO_2 and H_2O, releasing the energy trapped within the chemical bonds. The energy released during oxidation of energy rich compounds is made available for activities of cells through an intermediate compound called Adenosine Triphosphate. (ATP – biological energy currency).

The breaking of complex organic substances through oxidation releases the energy and this process is called respiration.

OR

Respiration is defined as an intracellular process of oxidation in which complex organic substances are broken down in step wise manner with the release of energy which is immediately converted into metabolically usable form of energy i.e. ATP.

The respiration may be defined as "Process of biological oxidation in living cells in which complex food materials are broken down stepwise into simple substances and to release energy".

$$C_6H_{12}O_6 + 6O_2 \xrightarrow[\text{oxidation}]{\text{Living cells with}} 6CO_2 + 6H_2O$$

$$+ \text{ Energy (38 ATP} - 686 \text{ kcal)}$$

3.2 TYPES OF RESPIRATION

On the basis of availability of oxygen during respiration, there are two main types of respiration:

(1) Aerobic respiration and

(2) Anaerobic respiration.

(1) Aerobic respiration: When respiration takes place in the presence of free or molecular oxygen and substrates for a respiration are completely oxidised into CO_2 and H_2O as byproduct and large amount of ATP energy is released as shown in following reactions.

$$C_6H_{12}O_6 + 6O_2 \longrightarrow 6CO_2 + 6H_2O + 38 \text{ ATP}$$

$$(686 \text{ kcal energy})$$

The aerobic respiration takes place in cytosol and mitochondria. It occurs in all living plants and animal cells.

(2) Anaerobic respiration: When respiration takes place in absence of free or molecular oxygen and there is incomplete oxidation of substrate, producing CO_2 and ethyl alcohol or lactic acids as byproducts and there is release of comparatively less ATP energy.

The equation is as follows:

$$C_6H_{12}O_6 \longrightarrow 2C_2H_5OH + 2CO_2 + 2 \text{ ATP}$$

$$(28 \text{ kcal energy})$$

Since many organisms are able to break down carbohydrates to CO_2 and some other substances such as alcohol, lactic acid etc.

without using molecular oxygen (O_2). This type of respiration is known as anaerobic respiration or fermentation because it is so often controlled by microorganisms or their enzymes.

The anaerobic respiration takes place in cytosol and also called as fermentation process. It occurs in micro-organisms like yeast, bacteria and fungi etc.

Mechanism of Aerobic Respiration:

Aerobic respiration is common to all plants and animals (eukaryotic organisms). It is complete oxidation of substrate, release lot of usable energy that can be readily utilized for various metabolic activities in cells.

The aerobic respiration takes place in the following steps:

(1) Glycolysis

(2) Formation of Acetyl Co-A

(3) TCA cycle.

(4) ETS in mitochondria

3.3 GLYCOLYSIS (EMP – PATHWAY)

The term Glycolysis is used to describe the sequential series of reactions present in a wide variety of tissues that starts with a hexose sugar (usually glocose) and ends with pyruvic acid. Glycolysis is defined as the stepwise, enzymatic breakdown of the hexose sugar (glucose) in to two molecules of pyruvate in cytoplasm. This term Glycolysis originated from Greek words Glycos means "glucose" or "sugar" and lysis means "splitting" or "break down". The scheme of glycolysis was discovered by three German scientists, namely Profs. Embden, Meyerhof and Parnas. Therefore glycolysis is also called as EMP-pathway.

Glycolysis occurs in all organisms and it is considered as most primitive form of carbon catabolism. The energy released during glycolysis is relatively low, but can be used to support-growth in some anaerobic organisms and in some tissues in aerobic systems under anaerobic conditions.

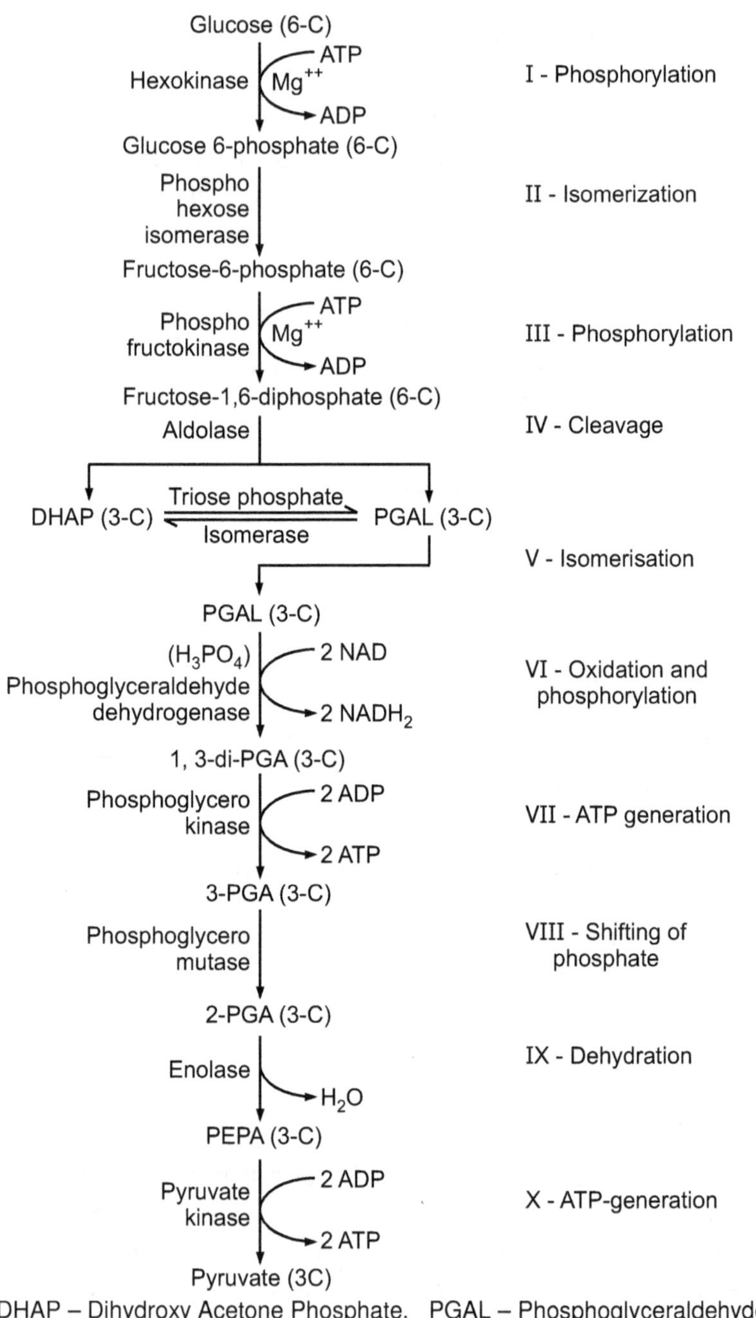

Fig. 3.1: Schematic representation of Glycolysis (EMP-Pathway)

Glycolysis consists of two main phases:

(A) Preparatory phase and cleavage (Step 1 to 5).

(B) Oxidative and payoff phase (Steps 6 to 10).

(A) Preparatory phase and cleavage:

It is initial phase of glycolysis in which glucose molecule is activated by phosphorylation and then cleaved into two molecules of triose – phosphates viz. 3-PGAL and DHAP. In the next step only 3 PGAL participates and therefore DHAP gets converted into 3-PGAL (two molecules) and it includes following reactions.

(I) Phosphorylation: (i.e. 1st – phosphorylation) : Glucose (6C) molecule undergoes phosphorylation with the help of ATP. It occurs in the presence of the enzyme hexokinase and co-factor Mg^{++}. In this reaction one phosphate group from ATP is removed to produce Glucose-6-phosphate (6C) and ADP.

$$\text{Glucose} + \text{ATP} \xrightarrow[\text{Mg}^{++}]{\text{Hexokinase}} \text{Glucose-6-phosphate} + \text{ADP}$$
$$\text{(6C)} \qquad\qquad\qquad\qquad\qquad \text{(6C)}$$

(II) Isomerisation: Glucose-6-phosphate is converted into fructose-6-phosphate in presence of enzyme phosphohexose isomerase and the co-factor Mg^{++}.

$$\text{Glucose-6-phosphate} \xrightleftharpoons[\text{Mg}^{++}]{\substack{\text{Phosphohexose}\\ \text{isomerase}}} \text{Fructose-6-phosphate}$$

(III) Phosphorylation: (i.e. IInd phosphorylation): Fructose-6-phosphate is phosphorylated (6C) is further phosphorylated by ATP molecule to give fructose 1, 6 diphosphate and ADP. Thus second phosphorylation takes place and phosphate group is attached to the first carbon. This reaction is catalysed by the enzyme phosphofructo kinase and the co-factor Mg^{++}.

$$\text{Fructose-6-phosphate} + \text{ATP} \xrightarrow{\substack{\text{Phosphofructo}\\ \text{kinase and Mg}^{++}}} \substack{\text{Fructose}\\ \text{1,6 diphosphate}}$$
$$\text{(6C)}$$
$$+ \text{ADP}$$

Above mentioned three steps are collectively described as the preparatory phase of glycolysis.

(IV) Cleavage (Splitting): Fructose 1,6 diphosphate splits into two molecules of Triose phosphate i.e. 3 PGAL (3 phosphoglyceraldehyde) and DHAP (Dihydroxyacetone phosphate) in presence of enzyme Aldolase which are interconvertible.

$$\text{Frucost 1,6 diphosphate} \xrightarrow{\text{Aldolase}} \text{3 phosphoglyceraldehyde}$$
$$\text{(6C)} \qquad\qquad\qquad\qquad \text{(PGAL)}$$
$$+$$
$$\text{Dihydroxyacetone phosphate (DHAP)}$$

(V) Isomerisation: 3 PGAL and DHAP are isomers of each other and are interconvertible. DHAP is converted into 3 PGAL, with the help of enzyme Triose phosphate isomerase.

$$\text{DHAP (3C)} \underset{\text{isomerase}}{\overset{\text{Triose phosphate}}{\rightleftharpoons}} \text{3 PGAL (3C)}$$

(B) Oxidative and pay off phase:

During this phase oxidation by removal of hydrogen takes place and then there is ATP generation (Harvest of energy).

(VI) Oxidation and Phosphorylation: In this reaction 3 PGAL oxidised to 1,3 diphosphoglyceric acid with oxidizing agent as NAD in presence of the enzyme phosphoglyceraldehyde dehydrogenease. Here two electrons (e^-) and two photons (H^+) which are given out by 3-PGAL reduce NAD to $NADH_2$. The reaction is accompanied by the addition of H_3PO_4 (phosphoric acid).

$$\text{3 PGAL} + H_3PO_4 + \text{NAD} \xrightarrow{\substack{\text{Phosphoglyceraldehyde} \\ \text{dehydrogenase}}}$$
$$\text{1,3 di-PGAL} + NADH_2$$

(VII) ATP-generation/formation: 1, 3-di-PGA is converted into 3-PGA by removal of a phosphate group. This phosphate combines with ADP to produce an ATP molecule.

The reaction takes place in presence of the enzyme phosphoglycerokinase.

$$1, 3 \text{ di-PGA} + ADP \xrightarrow{\text{Phosphoglycero kinase}} 3PGA + ATP$$

(VIII) Shifting of enzyme: 3-phosphoglycerate is converted into its isomer 2-phosphoglycerate by catalytic activity of enzyme phosphoglyceromutase. In this reaction the phosphate group is shifted from third carbon atom to the second carbon atom.

$$3 \text{ PGA} \underset{}{\overset{\text{Phosphoglyceromutase}}{\rightleftharpoons}} 2 \text{ PGA}$$

(IX) Dehydration: 2PGA undergoes dehydration of H_2O and formation of PEPA (phospho-enol pyruvic acid) in presence of enzyme enolase.

$$2 \text{ PGA} \xrightarrow{\text{Enolase}} PEPA + H_2O$$

(X) ATP-generation: Phospho-enol pyruvic acid losses a phosphate group and gives pyruvic acid or pyruvate. The released phosphate group combines with ADP and formation of ATP in presence of the enzyme pyruvate kinase.

$$PEPA + ADP \xrightarrow{\text{Pyruvate kinase}} Pyruvate + ATP$$

The overall equation of glycolysis may be written as –

$$\text{Glucose} + 2 \text{ NAD} + 2 \text{ ADP} + 2 \text{ Pi} \xrightarrow[\text{Glycolysis}]{\text{Enzyme}}$$

$$2 \text{ pyruvate} + 2 \text{ NADH}_2 + 2ATP$$

Generation and utilization of ATP during Glycolysis:

During glycolytic pathway, the molecules of ATP are produced as follows:

(i) Direct transfer of phosphate to ATP.

(ii) Oxidation of NADH produced during glycolytic pathway to NAD.

In the end of glycolysis, Net gain of ATP:

(1) PGAL to 1,3 di-PGA = 2 NAD × 03   = 06 ATP

(2) 1,3 di-PGA to 3 PGA   = 02 ATP

(3) Phosphoenol pyruvic to pyruvic acid   = 02 ATP

　　　　　In EMP pathway synthesis of   = 10 ATP

But 2 ATP is lossed in between the reaction number I and III.

(1) Glucose to Glucose-6-phosphate   = 01 ATP

(2) Fructose-6-phosphate to fructose 1,6 diphosphate = 01 ATP

　　　　　　　　　　　　　　　　　　　　　　02 ATP

　　　　　　　　　　　　　　　　　　　　= 08 ATP

is the net gain in the glycolysis/EMP pathway.

3.4 FORMATION OF ACETYL CO-A

(Decarboxylation: Conversion of pyruvate to Acetyl-Co-A): Pyruvic acid is synthesized by glycolysis in cytosol. The TCA required for further oxidation of pyruvic acid are present in matrix space of mitochondrion. Therefore, for further oxidation pyruvic acid must be translocated from cytosol to a mitochondrion. This translocation takes place from cytosol through membranes of mitochondrial envelope to the matrix.

The oxidative decarboxylation of pyruvate to Acetyl Co-A involves the presence of at least five essential co-factors and a complex enzymes. The co-factors involved are:

(1) Mg ions.

(2) Lipoic acid (oxidized)

(3) Thiamine pyrophosphate (TPP)

(4) Co-enzymes – NAD

(5) Co-enzymes – A (Co-A).

Different steps in the oxidative decarboxylation of pyruvic acid to Acetyl Co-A are summarised as follows:

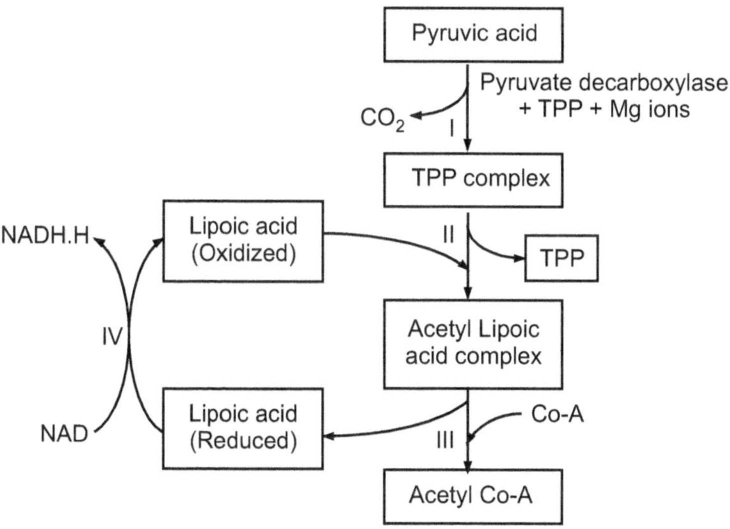

Fig. 3.2: Formation of Acetyl-Co-A from pyruvic acid

The reaction taking place in the oxidative decarboxylation of pyruvic acid are shown below:

(I) Pyruvic acid + Thiamine pyrophosphate (TPP) $\xrightarrow[\text{+ Mg ions}]{\text{Pyruvic dehydrogenase complex}}$ TPP Complex + CO$_2$↑

(II) Thiamine pyrophosphate (TPP) complex + Lipoic acid (oxidized) $\xrightarrow[\text{complex}]{\text{Pyruvic dehydrogenase}}$ Acethyl Lipoic acid complex + TPP complex

(III) Acetyl Lipoic acid complex + Co-A $\xrightarrow[\text{complex}]{\text{Pyruvic dehydrogenase}}$ Acetyl Co-A + Lipoic acid (reduced)

(IV) Lipoic acid (reduced) + NAD $\xrightarrow[\text{complex}]{\text{Pyruvic dehydrogenase}}$ Lipoic acid (oxidized) + NADH · H

The summary representation of the reactions.

$$\text{Pyruvic acid} + \text{NAD} + \text{Co-A} \xrightarrow{\text{Pyruvic dehydrogenase complex}}$$

$$\text{Acetyl Co-A} + CO_2 + \text{NADH} \cdot \text{H}$$

Acetyl group is finally linked to sulphur protein coenzyme-A (Co-A). In this process, NAD is reduced to $NADH_2$. In the form of Acetyl Co-A, the original pyruvic acid is further oxidized by TCA cycle reactions in a mitochondrion. The reduced co-enzyme $NADH_2$, is utilized for oxidation reactions. Atmospheric O_2 also utilized with the help of ETS in mitochondrion. The released energy in this oxidation is conserved in ATP molecules.

At the end of decarboxylation, from original glucose molecule two molecules of Acetyl-Co-A, two molecules of CO_2, four molecules of $NAD \cdot H_2$ are produced.

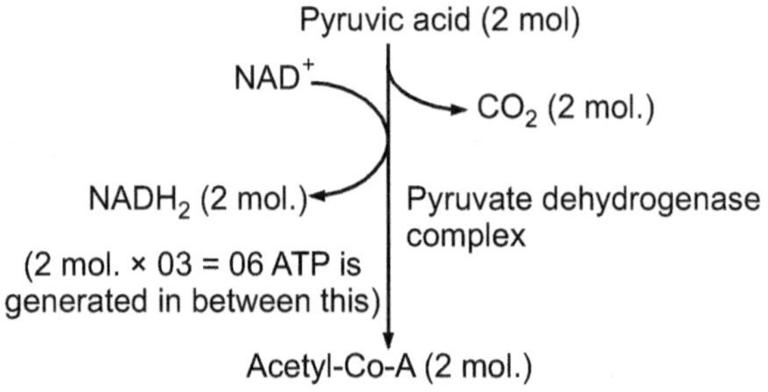

Fig. 3.3

It is the process of decarboxylation and dehydrogenation hence called as ***oxidative decarboxylation of pyruvic acid***. In this mechanism conversion of pyruvic acid to Acetyl-Co-A NAD is converted into $NADH_2$ (2 mol.) means 06 ATP is generated in this mechanism.

3.5 TCA CYCLE (KREBS CYCLE)

The first stable compound of Krebs cycle is a citric acid or citrate. Chemically it is a tri-carboxylic acid hence, it is called as Tri-carboxylic acid cycle or citric acid cycle. Various reactions of TCA cycle were discovered by Sir Hans Adolf krebs in 1937. Therefore, it is popularly known as Krebs cycle. The discovery of this pathway opened a new chapter in cellular biochemistry. Hence, Sir. Hans Adolf krebs was awarded Nobel Prize in Biochemistry in 1953/54.

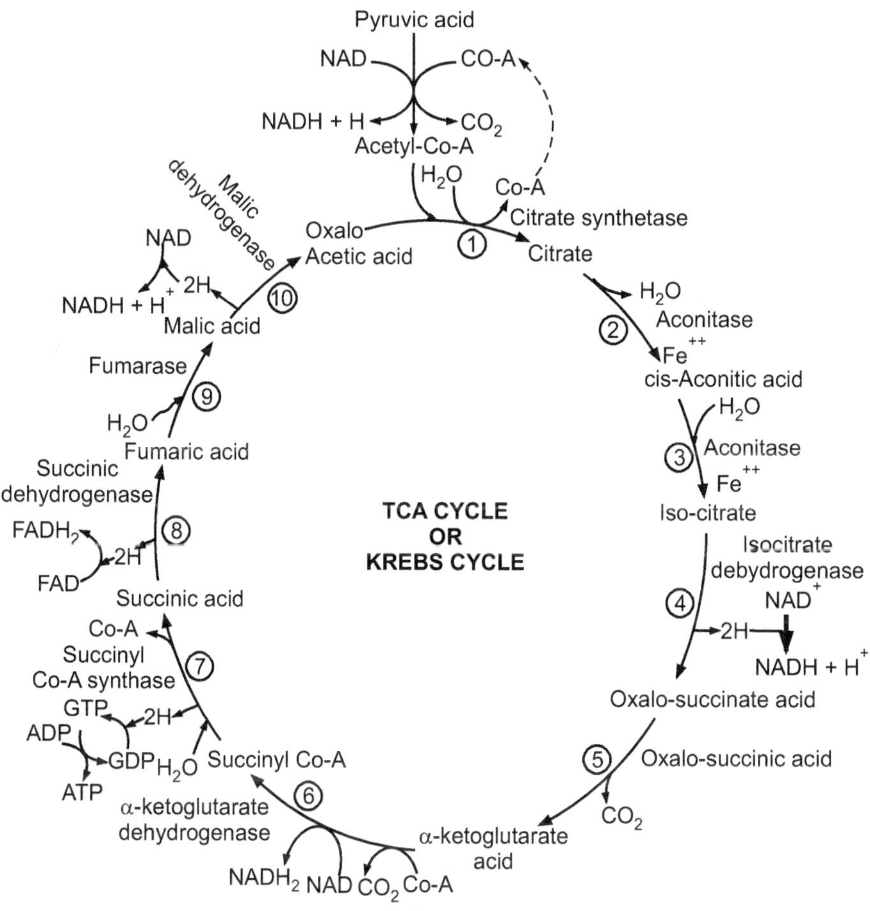

GDP – Gaunine Diphosphate

GTP – Gaunine Triphosphate

FAD – Flavin Adenosine Diphosphate

FAM – Flavin Adenosine Monophosphate

NAD – Nicotine Amide Dinucleotidic

Co-A – Co-enzyme-A

Fig. 2.4: TCA Cycle/Krebs Cycle/CAC Cycle

The cyclic process through which Acetyl-Co-A is completely oxidized and CO_2 is released in stepwise manner. It is second phase in the mechanism of aerobic respiration. It occurs in the mechanism of aerobic respiration. It occurs in the matrix of mitochondria, in the presence of several enzymes and co-enzymes. During this phase, co-enzymes NAD and FAD take up the $2H^+$ removal during oxidation steps and $NADH_2$ and $FADH_2$ are formed. It involves a series of reactions, each catalysed by a specific enzymes.

These Acetyl Co-A molecules are utilized for Krebs cycle reactions and not the pyruvic acid directly. TCA cycle shows three phases (1) condensation phase, (2) Oxidation phase, (3) Regeneration phase.

In the first reaction of Krebs cycle, one molecule of acetyl-Co-A combines with 4-carbon oxaloacetic acid (OAA); the result-6-carbon citric acid/citrate is produced and Co-A is released in presence of water and enzyme citrate synthatase.

$$\text{Acetyl Co-A + Oxalo Acetic Acid + } H_2O \xrightarrow{\text{Citrate synthatase}}$$

$$\text{Citrate + Co-A}$$

Net ATP gain from glucose in Krebs Cycle:

Sr. No.	Steps in Krebs Cycle	ATP Generated
1.	2 isocitrate $\longrightarrow$ Oxalosuccinate acid + $2NADH_2$ = $(2NADH_2 = 2 \times 3$ ATP$)$	= 06 ATP
2.	α-ketoglutarate $\longrightarrow$ Succinyl Co-A + $2NADH_2$ $(2NADH_2 = 2 \times 3$ ATP$)$	= 06 ATP
3.	2-Succinyl-Co-A $\longrightarrow$ 2 Succinic acid + 2GTP. $(2$ GTP $= 2$ ATP$)$	= 02 ATP
4.	Succinic acid $\longrightarrow$ 2 Fumaric acid + $FADH_2$ $2 FADH_2 = (2 \times 2$ ATP$)$	= 04 ATP
5.	2 Malic acid $\longrightarrow$ Oxaloacetic acid + 2 $NADH_2$ $2NADH_2 = 2 \times 3$ ATP	= 06 ATP
	Net gain of ATPs	**= 24 ATP**

Major reactions of Krebs cycle are as follows:

(1) Condensation: Acetyl Co-A combines with OAA which is a 4-C. Compound present in matrix of mitochondrion and a 6-C. Compound called citric acid or citrate is formed. Citrate is the first stable compound of Krebs cycle. One water molecule is used in this reaction. The enzyme citrate synthetase catalyses the reaction. Here, Co-A is released for recycling.

$$\underset{(2C)}{\text{Acetyl Co-A}} + \underset{(4C)}{\text{OAA}} + H_2O \xrightarrow{\text{Citrate synthatase}} \underset{(6C)}{\text{Citrate}} + \text{Co-A}$$

Citric acid is a six carbon tri-carboxylic acid. Hence, gives the name TCA cycle. The above reactions is the condensation reaction.

(2) Isomerisation: Citric acid gets converted into its isomers, isocitrate in the presence of enzyme aconitase and Fe^{++}.

Dehydration: Citrate losses a water molecule and formation of cis-aconitic acid in presence of enzyme aconitase and Fe^{++}.

$$\text{Citrate} \xrightarrow[Fe^{++}]{\text{Aconitase}} \text{Cis-Aconitic acid} + H_2O$$

(3) Hydration: Cis-aconitic acid combines with a water molecules and formation of iso-citrate in presence of enzyme aconitase and Fe^{++}.

$$\text{Cis-aconitic acid} + H_2O \xrightarrow[Fe^{++}]{\text{Aconitase}} \text{Iso-citrate}$$

(4) Oxidation (Dehydrogenation): Iso-citrate is oxidized to form oxalo-succinate, two hydrogen atoms released at this step are taken up by the co-enzyme NAD and $NADH_2$ is formed in presence of enzyme iso-citrate dehydrogenase and Mn^{++}.

$$\text{Iso-citrtae} + \text{NAD} \xrightarrow[Mn^{++}]{\text{Iso-citrate dehydrogenase}} \underset{\text{succinate}}{\text{Oxalo}} + NADH_2$$

(5) Decarboxylation-I: Oxalo - succinate undergoes decarboxylation and formation of α-ketoglutarate acid in presence of enzyme oxalo succinic acid and CO_2 goes out of the mitochondria.

$$\text{Oxalo-succinate acid} \xrightarrow{\text{Oxalo succinic acid}} \alpha\text{-ketoglutarate acid} + CO_2\uparrow$$

(6) Oxidative: (Decarboxylation-II and Dehydrogenation-II): α-ketoglutarate acid undergoes oxidative decarboxylation i.e. oxidation (dehydrogenation) and decarboxylation. During this, CO_2 is released hydrogen removed from the substrate is received by the co-enzyme NAD to form $NADH_2$. Succinyl Co-A is formed in presence of enzyme α-ketoglutarate dehydrogenase.

$$\alpha\text{-ketoglutarate} + NAD + Co\text{-A} \xrightarrow{\alpha\text{-ketoglutarate dehydrogenase}} \text{Succinyl Co-A}$$

$$+ NADH_2 + CO_2$$

(7) Hydration and phosphorylation: Succinyl Co-A gets hydrated using water molecules and formation of succinic acid, in presence of enzyme succinyl Co-A synthatase. During this reactions energy liberated and used in the formation of GTP from GDP and Pi. However, GTP is unstable as compared to ATP. It is soon transformed to ATP using ADP.

$$\text{Succinyl Co-A} + H_2O + GDP \xrightarrow{\text{Succinyl-Co-A synthatase}} \text{Succinic acid}$$

$$+ GTP + Co\text{-A}$$

$$GTP + ADP \longrightarrow GDP + ATP \longrightarrow \text{Substrate level phosphorylation}$$

(8) Oxidation (Dehydrogenation – III): It is third oxidative step in krebs cycle. The enzyme succinic dehydrogenase is only oxidative enzyme requiring co-enzyme, FAD. The reduced co-enzyme-$FADH_2$ is added in mitochondrial matrix for using O_2 with the help of ETS.

$$\text{Succinic acid} + FAD \xrightarrow{\text{Succinic dehydrogenase}} \text{Fumaric acid} + FADH_2$$

(9) Hydration: Fumaric acid accept a water molecule and to produce malic acid in presence of enzyme fumarase.

$$\text{Fumaric acid} + H_2O \xrightarrow{\text{Fumarase}} \text{Malic acid}$$

(10) Oxidation (Dehydrogenation – IV): Malic acid is oxidized by removal of hydrogen and Oxalo Acetic Acid gets regenerated. The NAD to form $NADH_2$. The reaction is catalysed by the enzyme malate dehydrogenase.

$$\text{Malic acid} + NAD \xrightarrow{\text{Malic dehydrogenase}} \text{Oxalo Acetic acid} + NADH_2$$

During krebs cycle, three molecules of NAD^+ and one molecule of FAD are reduced to produce NADH and $FADH_2$, respectively. In the krebs cycle, NADH and $FADH_2$ are produced. Now, they are linked with ETS and produce ATP by oxidative phosphorylation. It may be summarised in the following equation.

$$\text{Pyruvic acid} + 4\, NAD + FAD + H_2O + ADP + iP \xrightarrow{\text{Mitochondrial matrix}}$$

$$3\, CO_2 + 4\, NADH + 4H^+ + FADH_2 + ATP$$

In the end of Krebs cycle, glucose molecule and two pyruvic acid molecules are formed. After oxidation of one pyruvic acid molecules are formed. After oxidation of one pyruvic acid molecules, three CO_2 molecules are released. Thus, in all six molecules of CO_2 are released.

Net gain of ATP in Aerobic Respiration:

(1) Glycolysis   = 08 ATP

(2) Decarboxylation   = 06 ATP (conversion of pyruvate to Acetyl – CO-A)

(3) TCA cycle   = 24 ATP

38 ATP Net gain of ATP

$$C_6H_{12}O_6 + 6\, O_2 \xrightarrow{\text{Glycolysis, Decarboxylation Krebs cycle}} 6\, CO_2 + 6H_2O + \text{Energy (38 ATP)}$$

3.6 ETS IN MITOCHONDRIA

Electron Transport System or Respiratory Chain: The oxidation of reduced coenzymes into their oxidized forms is known as terminal oxidation. The metabolic pathway through which electron passes from one carrier to another carrier is called the Electron Transport System (ETS). The electron transport system is also known as 'electron transport chain' or 'mitochondrial respiratory chain'. The ETS consists of a series of co-enzymes and cytochromes that take part in passage of electrons from a chemical to its ultimate acceptor. Free molecular oxygen is the final acceptor of electrons. In this process, synthesis of ATP takes place with the help of energy released during the electron transfer. This synthesis of ATP is called oxidative phosphorylation.

At the time of glycolysis and krebs cycle, respiratory substrates are oxidized at several steps. It occurs dehydrogenation, this hydrogen is accepted by co-enzymes – mostly NAD and sometimes FAD. So that they are reduced to $NADH_2$ and $FADH_2$ respectively. Such reduced co-enzymes are converted back to their oxidized forms. During the process the H_2O removed from reduced co-enzymes is released into the matrix of mitochondria. The hydrogen reacts with atmospheric oxygen to form water molecule. This is called metabolic water and is formed at the end of electron transport system (ETS), and hence is called terminal oxidation.

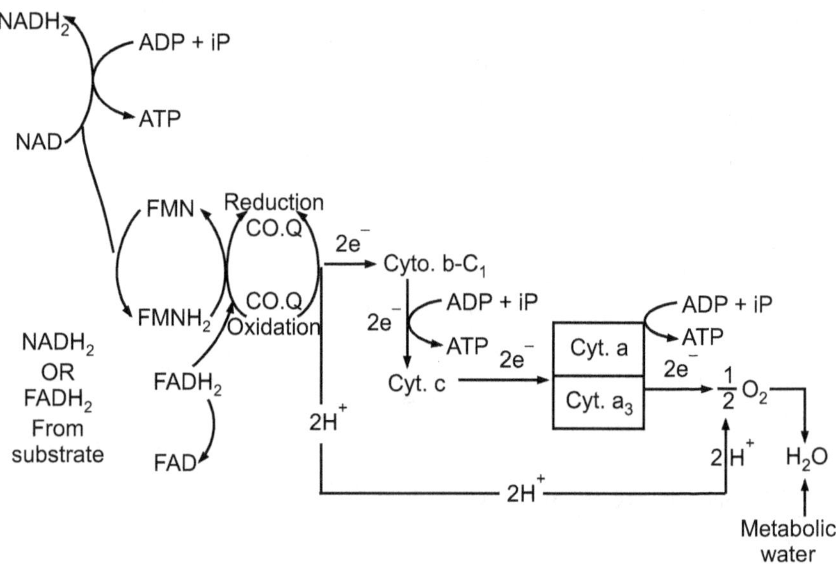

Fig. 3.5: Electron Transport System (Terminal Oxidation)

ETS cycle includes FMN as the co-enzyme, co-enzyme-Q (ubiquinone) as both, a coenzyme and an electron carrier and a number of electron carriers called cytochrome. They include cytochromes b, C_1, c, a and a_3. Cytochromes are iron containing compounds. Iron keeps oscillating between ferric and ferrous forms when electrons ($2e^-$) are taken up by the cytochrome, ferric compound gets reduced to ferrous and as it gives the electrons to the next cytochrome in the chain it gets reoxidized to ferric. The final electron acceptor is atmospheric oxygen.

Different electron carriers are arranged on the body of the oxysomes in the order of their decreasing energy level. NAD is the first and O_2 is the last energy level. So NAD is the highest energy level and oxygen is the lowest-energy level. When a reduced co-enzyme $NADH_2$ or $FADH_2$ from glycolysis or Kreb cycle, enters the respiratory chain, hydrogen is removed and NAD and FAD is regenerated. The hydrogen further split into proton and an electron. (H^+ and e^-)

$$H_2 \longrightarrow 2H^+ + 2e^-$$

The electrons pass through the electron transport system so that the electron acceptors get oxidized and reduced alternately. They move along cytochrome b, C_1, c, a and a_3. The protons are released in the mitochondrial matrix. Finally, the two electrons react with atmospheric oxygen to produce $O^{- -}$ which joins with $2H^+$ and produce a water molecule. This is called metabolic water.

Electrons transported with electron carriers, and energy released in each transfer. At certain stages, it is utilized for synthesis of ATP from ADP and Pi. The biosynthesis of ATP molecule using energy released during oxidation of respiratory substrates is called oxidative phosphorylation.

3 ATP molecules are synthesized through ETS during electron transfer. The first ATP molecule is formed between $NADH_2$ and FMN, second ATP between cytochrome complex – bc_1 and cytochrome - c and third ATP between cytochrome a and cytochrome a_3. Thus, when $NADH_2$ is reoxidized, three ATP molecules are formed and when $FADH_2$ is reoxidized, only two ATP molecules are formed.

3.7 FERMENTATION

Fermentation is defined as an "anaerobic breakdown of carbohydrates and other organic compounds, into alcohol, organic acids, gases etc. with the help of microorganisms or their enzymes". In micro-organisms, such as bacteria and yeasts, the term 'Fermentation' is more commonly used where anaerobic respiration takes place, after the name of product like alcoholic fermentation, lactic acid fermentation, etc. The enzyme complex present in the process was named as 'zymase'.

Scientist Buchner in 1897 found that, the extract of yeast cells (which does not contain living cells) can bring about fermentation of sugar solution. It was discovered that the enzyme zymase present in the yeast extract is capable of fermenting activity i.e. breakdown of carbohydrates together with the formation of alcohol and CO_2.

The mechanism of anaerobic respiration or fermentation is similar to common pathway of aerobic respiration up to glycolysis. However, oxidation of pyruvic acid and NADH by mitochondria requires oxygen, plants carry out fermentation (anaerobic respiration) which leads to the formation of CO_2 and ethanol (ethyl alcohol) or lactic acid. In most of the cases, ethanol is produced.

Two Types of fermentation:

1. Alcoholic fermentation: In this type, the pyruvic acid is converted into Acetyldehyde in the presence of enzyme carboxylase. Thiamine pyro-phosphate (TPP) is required as a co-factor.

$$2\ CH_3COCOOH \xrightarrow[\text{TPP}]{\text{Carboxylase}} 2CH_3CHO\ +\ 2CO_2 \text{ (Acetyldehyde)}$$

Glycolysis:

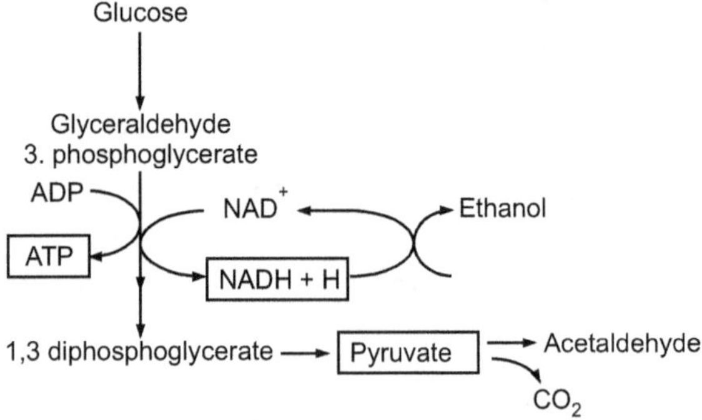

Fig. 3.6: Pathway of fermentation in yeast (anaerobic respiration)

Acetyldehyde is then reduced to ethanol (ethyl alcohol) by the enzyme alcohol dehydrogenase co-enzyme-$NADH_2$ (produced in glycolysis) is oxidised.

$$2CH_3CHO + NADH_2 \longrightarrow CH_3CH_2OH + NAD$$

(Acetyldehyde) (Reduced)　　　　(Ethanol) (Oxidized)

The overall equation for anaerobic respiration involving alcoholic fermentation:

$$C_6H_{12}O_6 \longrightarrow 2\ CH_3CH_2OH + 2\ CO_2$$

Alcohol

2. Lactic Acid Formation: Here, Pyruvic acid and co-enzyme $NADH_2$ is oxidised and formation of lactic acid and NAD in presence of enzyme lactic dehydrogenase.

$$CH_3CO \cdot COOH + NADH_2 \xrightarrow[\text{dehydrogenase}]{\text{Lactic}} CH_3CHOH \cdot COOH + NAD$$

Pyruvic acid　　　　　　　　　　　　　　Lactic acid

The overall equation for anaerobic respiration (Lactic acid fermentation) is as follows:

$$C_6H_{12}O_6 \longrightarrow 2\ CH_3CHOH \cdot COOH$$

Glucose　　　　　　　Lactic Acid

There is a net gain of only 2 ATP molecules (in glycolysis) during anaerobic respiration (or fermentation) and most of energy contained in glucose molecule is released as heat.

Importance:

Fermentation processes are of great importance in the household and industry. An example of household is formation of curd from milk. In the industry-the alcohol and acid fermentation. The ripening of cheese, the curring of tobacco and tea leaves, sum hemp, ramie and jute, tanning of leather are very important examples of commercial processes dependent upon fermentations brought about by different species of bacteria. Alcoholic fermentation is used in brewing industry, where various types of beers, whisky and wines are produced, whereas CO_2 of alcoholic fermentation is used in baking industry for making bread and biscuits. Vinegar is produced by fermentation activity of acetic acid bacteria. Clearing of raw hides is done by fermentive activity of bacteria.

QUESTIONS

A. Multiple Choice Questions:

1. Respiration occurs in ....
   (a) all living cells in light
   (b) all living cells in dark
   (c) both (a) and (b)
   (d) non-green cells in dark only.

2. Respiration may occur in absence of O_2 in ....
   (a) potato   (b) spirogyra
   (c) yeast   (d) humans

3. Kerbs cycle is also called as ....
   (a) Glycolate cycle   (b) EMP pathway
   (c) Glyoxylate cycle   (d) Citric acid cycle

4. The net gain of energy from one gram mole of glucose during aerobic respiration is ....
   (a) 38 ATP   (b) 40 ATP
   (c) 36 ATP   (d) 2 ATP

5. Respiratory enzymes are located in ....
   (a) matrix of mitochondria   (b) perimitochondrial space
   (c) cristae   (d) outer membrane

6. ATP synthesis occurs in ....
   (a) mitochondria as a whole
   (b) inner membrane of mitochondria
   (c) outer membrane of mitochondria
   (d) mitochondrial matrix.

7. Which one is the product of glycolysis?
   (a) FAD   (b) NAD
   (c) NADP   (d) NADH

8. Kerbs cycle is a part of .... respiration.
   (a) Aerobic   (b) Anaerobic
   (c) Fermentation   (d) Facultative

9. Respiration is a process in which ....
   (a) energy is used up   (b) energy is released
   (c) energy is released and stored in the form of ATP
   (d) energy is stored in the form of ADP
10. End products of respiration in plants are ....
   (a) Starch and O_2 (b) H_2O and energy
   (c) CO_2, H_2O and energy   (d) Sugar and O_2
11. Glycolysis takes place in ....
   (a) Cytoplasm   (b) Chloroplast
   (c) Ribosome   (d) Mitochondria
12. Total ATP production during EMP pathway is ....
   (a) 24 ATP   (b) 8 ATP
   (c) 38 ATP   (d) 6 ATP
13. Kerbs cycle occurs in ....
   (a) endoplasmic reticulum   (b) chloroplast
   (c) mitochondria   (d) dictyosomes
14. Site of glycolysis or EMP is ....
   (a) cytoplasm   (b) mitochondrion
   (c) chloroplast   (d) nucleus
15. Electron acceptor in cyt a_3 is ....
   (a) Fe   (b) Cu
   (c) As   (d) both (a) and (b)

> **Answer key:**
>
> (1) – c, (2) – c, (3) – d, (4) – a, (5) – a, (6) – b, (7) – d, (8) – a,
> (9) – c, (10) – c, (11) – a, (12) – b, (13) – c, (14) – a, (15) – a

B. Broad Questions:

1. Write an essay on glycolysis.
2. Give an account of Krebs cycle.
3. Define respiration. Explain the complete process of respiration in plants.
4. Describe the process and role of citric acid cycle in living organisms.

C. Write short notes on:

1. Glycolysis
2. Electron Transport Chain
3. Conversion of Pyruvate to Acetyl Co-A
4. Aerobic respiration
5. Fermentation
6. Types of respiration.

SEED DORMANCY AND GERMINATION

4.1 SEED DORMANCY

Definition of Seed:

"Seed is a matured and fertilized ovule". OR

According to Kozlowski and Gunn (1972) "A true seed is a matured and fertilized ovule that possesses an embryonic regions and a protective cover of the seeds".

Seed is the reproductive structure characteristic of all phanerogams (flowering plants). The structure of seeds may be studied in such common types of pea, bean, almond and sunflower.

4.2 CONCEPT OF DORMANCY

The growth of the seed is completely stopped after it is fully developed. In some cases growth is suspended because of the lack of favourable conditions in the environment. Such seeds will germinate if they are supplied with water and suitable temperature. In many other plant cases, the seeds would not germinate, even if they are provided with water etc. In such cases, the completely dry ripe seed in physiologically inactive and is said to be in a resting stage. The seed is called dormant and the phenomenon is termed as dormancy. If a seed does not germinate because of lack of one or more of the environmental factors, it is described as inactive. In a seed, the growth of the embryo is blocked during the period of dormancy. Different parts of the seed like seed coats, endosperm and embryo may be responsible for dormancy in the seed.

Seed dormancy is defined as "seeds are unable to germinate even under favourable environmental conditions". These conditions are a complex combination of water, light, temperature, gases, mechanical restrictions, seed coats and hormone structures. Dormant seeds are also called as resting stage quiescent seeds. (period of inactive seed).

Dormant seed shows following features:

Low moisture content, low turgor pressure in cell, insolube reserve food in endosperm, reduced enzyme activity and very low rate of aerobic respiration.

Generally, all plants experience a period of suspended growth. The suspension may be due to exogenous control. Such as a change in environmental conditions and is referred to as quiescence. In majority of the cases, true dormancy helps the seed to survive through an unfavourable season e.g. a very cold winter or a very hot summer.

4.3 CAUSES OF DORMANCY (FACTORS RESPONSIBLE FOR DORMANCY)

The dormancy of seed may be caused by the following factors:

(1) Hard Seed Coat:

A hard seed coat can cause dormancy in three ways:

(a) Seed coat impermeable to water: The seed coats of many species are completely impermeable to water at the time of their maturity. This condition is very common in the seeds of families like Leguminosae, malvaceae, chenopodiaceae, convolvulaceae and solanaceae have very hard seed coats. Such seed coats are impermeable to water.

(b) Seed coats impermeable to oxygen: In some species the hard seed coat is completely impermeable to oxygen. Such seeds would not germinate unless they have undergoes the period of dormancy during which the permeability of the seed coats will gradually increase. Two types of seeds present in the fruit of 'Cocklebur' (*Xanthium* species-Asteraceae) are not equally dormant. The lower (weak) seeds of 'bur' plants germinates under normal conditions of moisture and temperature while the upper (healthy) seeds would germinate only if it is kept under high oxygen conditions (tension) or its seed coat is removed. Sci. Wareing and foda (1956) believe that an inhibitors, present in the upper seed and it is oxidized by high oxygen tensions and germination is facilitated.

(c) Mechanically resistant seed coats: In some seeds of mustard, pigweed (*Amaranthus retroflexus* – Amaranthaceae), the seeds are permeable to water and oxygen but the resistant seed coat would not allow the embryo to development. The embryo of these seeds have no dormant period and will grow radily, if the seed coats are removed.

(2) Immature embryo:

In the certain species of orchidaceae, orobanchaceae, Ranunculaceae (*Ranunculus sps*) and *Ginkgo biloba* (Gymnosperm). The seeds are liberated, the embryo is not fully developed. Dormancy of this type is a natural consequence of the presence of an incompletely developed embryo should be allowed to complete development in an environment, which is favourable for germination.

(3) Period of after ripening:

Seed germinate after they have completed a period of rest, which is called as a "period of after ripening". Seeds of apple, peach, iris, tulip, poplar, pines and many other species could not germinate, even the conditions are favourable for germination. In certain species, the processes of after ripening is completed during the period of dry storage, while in others it is completed only in the presence of moisture and low temperatures and the condition is called as **"stratification"**. Opinion of the many workers is that in this period the embryo is present in a completely "dormant" stage. Dormancy of cereal seeds may be removed by staring them at 35 to 40°C for 2 – 4 days.

(4) Germination inhibitors:

Seeds of certain plant contain compounds which inhibit their germination. Such natural germination inhibitors have been found in the pulp of the juice of the fruit of family – Cucurbitaceae, Rutaceae, Asteraceae (example: xanthlum) and – solanaceae (example: Tomato) etc. containing seeds, seed coat, endosperm and embryo etc. Some of the inhibitors are ammonia, phthalids, coumarin, para-ascorbic acid, abscisic acid, organic acids, phenolics, alkaloids, aldehydes. Tomato pulp contains – caffeic acid and ferulic acid which are inhibitors of germination.

4.4 METHODS OF BREAKING OF SEED DORMANCY

Dormancy of seeds is very useful to man, but it is not linked by the farmers, who would like the seeds to germinate soon after they are harvested. A number of methods are employed for the breaking of dormancy. The employed methods differ from species to species depending up on the cause of dormancy:

(1) Scarification (removal of hard seed coat):

"Scarification is a treatment by which the hard seed coat is ruptured or weakened". The treatment may be mechanically and chemically.

By the treatments the expansion of the embryo is facilitated (make easy) and oxygen gain into the seed. The treatments may be mechanically, chemically and naturally.

(a) Mechanical scarification: Several types of mechanical scarification:

(1) Threashing of the seeds by machine.

(2) Rupturing the seeds with a hammer.

(3) Cutting the seed coat by the scalpel/blade etc.

(b) Chemical scarification: By keeping the seeds in strong acid like HCL, HNO_3 and H_2SO_4 for few seconds to some minutes or certain organic solvent like – Alcohol, xylene, chloroform and ether. The seeds kept in organic solvent for some time helps in breaking the dormancy.

(c) Natural scarification: The seeds remains for longer period in the environment, the dormancy is broken by the slow decaying of the hard seed coats due to the climatic activity i.e. Natural scarification.

(2) Low temperature (Chilling treatment):

In many herbaceous and woody plant species the dormancy can be broken by the chilling treatments (just above freezing –0°C to 05°C). Under field conditions this chilling requirement is normally met in the natural winter season, but such seeds do not germinate in the autumn season when they are shed and their germination is delayed until the following spring season.

It is noted that, in some species only intact seed which requires a chilling treatment (embryo germinates easily if seed coats are removed), in many other species the embryo shows dormancy, which has to be overcome by chilling.

The exact role of low temperature (chilling treatment) in breaking dormancy is not clearly understood. But some scientists believe that, low temperature has a favourable effect on the rate of respiration of the seeds.

(3) Alternating temperature:

The seeds are in the dormant stage, dormancy may be broken by exposing them alternately to low and high temperature. Care should be taken, the difference between the two extremes of temperature is not more than 10 to 20°C.

(4) Light:

The germination of many seeds is affected by light are termed as **photoblastic**. Seeds in which germination is stimulated by light are called *positively photoblastic*. Examples – pig weed – *Amaranthus retroflexus, Lactuaca sativa*, Tobacco – *Nicotiana tabaccum*, Tomato-*Lycopersicum esculantum* etc. Whereas those in which germaintion is inhibited by light are negatively photoblastic. Examples: *Nigella damascena, silene armeria, Nemophila sps.* etc.

Flint and **McAlister** (1935, 37) working on lettuce – *Lactuca sativa* (Asteraceae) seeds, found that the 'red light' (650 A° wavelength) was most effective in promoting the seed germination. Whereas 'far-red' light (730 A° wavelength) inhibits the seed germination. The pigments which brings about this, state of seedlings has been termed 'phytochromed'.

(5) Pressure:

It is strongly believed that, the pressure increases the permeability of the seed coats to water. This effect carry on even after the seeds have been dried and stored. Davies (1928) reported highly improved germination in the seeds of *sweet clover* (B.N. *Mililotus officinalis* – Fabaceae), and alfalfa (B.N – *Medicago sativa* – Fabaceae) they were subjected to hydraulic pressures at 2000 atm at 18°C. He also found an increase in the germination of the seeds by 50 to 200%, if the pressure was applied for 5 – 20 minutes.

(6) Hot water treatments:

Seeds soaked in hot water (77 to 100°C) container then seeds removed from hot water and allow to cool and soak for 24 hours.

Or - In this treatment the seeds are immersed in 40 – 60°C of hot water for 5 to 10 hours respectively then the seeds were incubated in glass jar for 24 hours, after incubation period the seeds are sown into beds.

(7) Treatment with growth regulators:

Use of low level of growth regulators (i.e. Gibberellins, Cytokinins and Ethylene etc.) may break the seed dormancy. Most commonly used growth regulators are gibberellins and kinetine. Pre-soaked seed treatment with GA_3 at the concentration of 100 ppm have been used for breaking seed dormancy.

4.5 SEED GERMINATION

Introduction and types: (Epigeal, Hypogeal and Viviparous)

4.5.1 Introduction

Seed germination is an important process in the life cycle of the plants. Seed is the connecting link between the two generations. Seed germination is the important process to continue the life of a plant.

Seed germination can be defined as "It is the resumption of active growth on the part of the embryo, resulting in the rupture of the seed coats and the emergence of the young plant is known as seed germination".

Seed germination is a complete biochemical process which covers all the steps from sprouting of the seed upto the development as an independent plant. The mature seed is live but the embryo is dormant. The low water contents (dryness) is the main cause of resisting the seed from germination. If the seed is viable and when the conditions like water, temperature, oxygen and light are favourable the seed immediately undergo some biochemical changes and germinates.

Structure of Seeds:

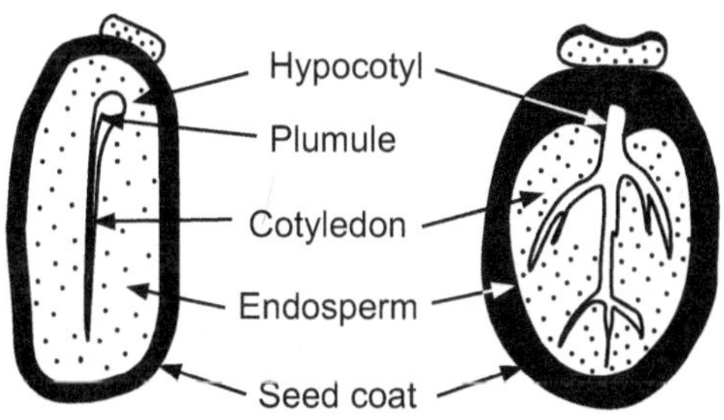

Fig. 4.1: Caster seed

4.5.2 Conditions Necessary for Seed Germination

i) **Water/moisture:** Water is the basic and most essential for seed germination. It makes seed coat soft. Water also helps to secrete digestive enzymes, which convert the complex food into simple forms.

ii) **Temperature:** Optimum temperature is essential for germination. The optimum temperature varies from plant to plant. The optimum temperature ranges from 25°C to 30°C.

iii) **O_2 supply:** A favourable O_2 level in the soil is essential for germination.

iv) **Light:** Light is essential for the seed germination.

4.5.3 Process of Seed Germination

It is very much important to understand the process of seed germination. The seed undergo some physical and bio-chemical changes during germination. These changes usually starts with the availability of water. These changes are : (1) Imbibition (2) Respiration (3) Effect of light (4) Use of reserve food material (5) Development of embryo axis into seedling.

1) Imbibition: Imbibition is the process where water enters the inner part of the seed and the seed gets swollen. Due to this the seed coat breaks and radicle come out to form root. Imbibition also cause the rehydration of structural or storage molecules.

2) Respiration: Imbibition of water causes the resumption of metabolic activity in the rehydrated seed. Initially their respiration is anaerobic but it soon becomes aerobic as O_2 enters the seed. The seeds of water plants, as also rice, can germinate under water by utilizing dissolved O_2. Seeds of land plants are sown on top surface of soil to get more O_2.

3) Effects of light on seed germination: Seeds are sensitive to light. Depending on the light requirement the seeds are called as:

(i)   **Positively photoblastic:** Need light for germination.

(ii) **Negatively photoblastic:** Light is not needed for germination.

(iii) **Non-photoblastic:** These seeds can germinate in both the conditions. The Red light is effective for germination.

The far-red light is harmful for the germination. Phytochromes regulates the light dependent process in plants.

4) Use of reserve food material: Seeds use the reserve food material stored in cotyledons.

5) Development of embryo axis into seedling: After the translocation of food and its assimilation, the cells of the embryo in the growing regions become metabolically very active. The cells grow in size and begin divisions to form the seedling.

4.5.4 Types of Seed Germination

Seeds germinates in the soil when all the conditions are favourable. Different seeds show different germination behaviours. The monocot and dicot seeds show variable germination behaviours. On the basis of relationship of the seed with the soil, the germination types are categorized as follows: (a) Hypogeal germination, (b) Epigeal germination and (c) Vivipary (Viviparous germination).

(1) Hypogeal germination:

This type of germination, the seed germinates but the cotyledons do not come out of the soil surface. During the germination, development, the epicotyl (embryonic axis between the plumule and cotyledons) show elongation due to which the plumule get pushed out of the soil. The plumule is elevated through the soil by rapid elongation of the epicotyl, which is the stem region between the cotyledons and the first true leaves or the first internode.

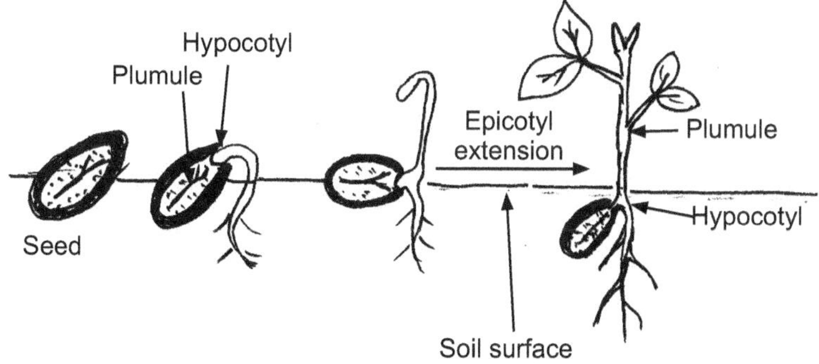

Fig. 4.2: Stages in Hypogeal germination

All monocotyledons show hypogeal germination. Some of the dicotyledons like gram, pea, groundnut show hypogeal germination. In monocots like wheat, maize, rice and coconut the radicle and plumule come out by breaking the coleorrhiza and coleptile respectively. The plumule grows upward and the first leaf comes out of the coleptile. The radicle forms the primary root which is replaced by many fibrous roots.

(2) Epigeal germination:

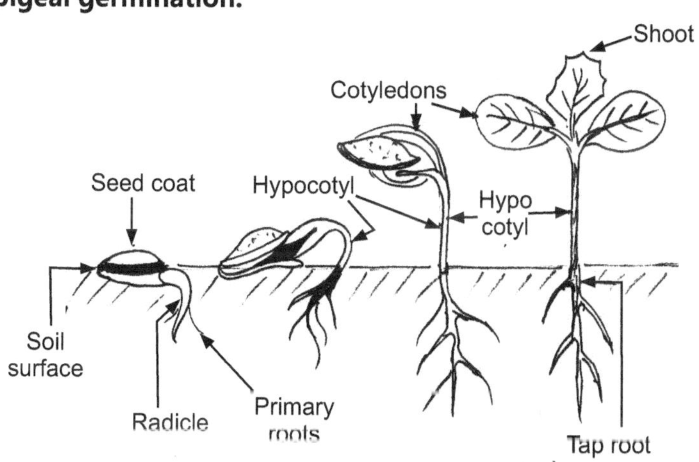

Fig. 4.3: Stages in epigeal germination

In this type of germination firstly the seed size increased due to imbibition (swelling). Due to continuous water absorption the seed coat breaks.

After rupture of seed coat the primary roots emerge from the lower end of hypocotyl or radicle. The primary roots are the first part of the embryo which is exposed to the external environment. The primary roots grow downward in the soil. It produce lateral roots and root hairs. The hypocotyl then elongates rapidly and pulls the cotyledons upward out of the soil into the air. After coming out of the soil both the cotyledons separate from each other in horizontal position on both the sides of the plumule. The plumule then begins active growth, which gives the stem and foliage leaves of the seedling. The food used during this process is from the cotyledons. Here, the cotyledons in addition to the food storage, it also perform photosynthesis till the seedling becomes independent. Very soon the cotyledons fall down as all the food material is used by the seedling. Many seeds show epigeal type of seed germination with some small changes e.g. Castor, Cotton, Papaya, Onion, Gourd.

(3) Viviparous seed germination:

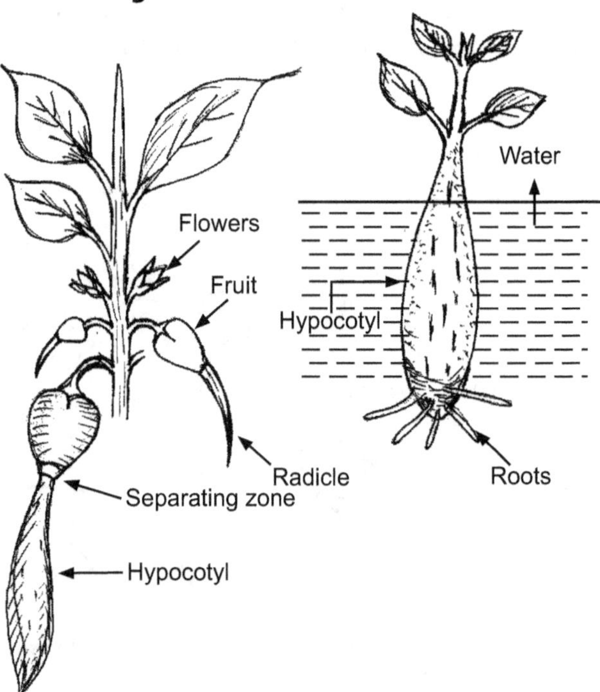

Fig. 4.4: Stages in viviparious seed germination in mangroves (e.g. *Rhizophora mucronata*)

Viviparous seed germination is a special type of germination or adaptation to avoid the harmful effects of saline conditions on seed germination. This type of seed germination is specifically found in saline water plants or mangroves. Mangroves usually grow in estuaries or salt marshy places where the saline water is available from tidal waters. The tidal water have high salt concentration (upto 35%) and high wave action. In such adverse conditions the delicate seeds fails to germinate and to produce a seedling. To escape from this situation the mangroves adopt for viviparous germination.

In this type of seed germination, the seeds or fruits germinate while attached to the mother plant. The seed germinates and produce seedling (in mangroves it is called propagule) on the mother plant. During this process the embryo of the seed present inside the fruit continues to grow while the fruit is attached to the mother plant. Hypocotyl elongates and pushes the radicle out of the seed and the fruit. Growth continues till the hypocotyl and radicle become several centimeters long (In *Rhizophora* it may reach 50 to 70 cm long). Due to such an extensive growth and weight the propagule become very heavy. The mature propagule show a yellow coloured separating zone in between the fruit and the propagule.

After maturation the propagule breaks its connection with the fruit and falls down in the salt rich muddy water. After falling in the mud the plumule remains outside the salt water while the pointed tip of the radicle gets fixed in the mud. This protects the plumule. The radicle quickly forms new roots and establishes the seedling as a new plant.

The viviparous seed germination show a seasonal adaptations. The germination of the seed and production of the propagule is generally takes place in summer season, when the temperature is high and salinity is also high. During this period the propagule become mature but never falls down. The mature propagule falls at the onset of monsoon, when the temperature is low and salinity of the water considerably reduced (as good as fresh water). Due to low salinity the propagules produce roots and hold the substratum tightly and show rapid growth.

The mangrove members of the family *Rhizophoraceae* show prominent development of viviparous seedling. The members are *Rhizophora mucronata, R.apiculata, Kandelia candel, Ceriops tagal, Brugiera gymorrhiza, Lumnitzera racemosa.*

The water softens the seed coat and helps to break the seed coat. It also increases permeability of seeds and converts the complex insoluble food into simple soluble forms for its availability to the developing embryo. Water makes the dissolved oxygen available for the growing embryo.

4.6 FACTORS AFFECTING SEED GERMINATION

Factors affecting seed germination can be categorized as:

(I)   External Factors

(II) Internal Factors

(I) External Factors:

(i) Water: Water is the most important external factor affecting germination. Absence of water in the seed make the seed dormant. Water is very much essential for imbibition, protoplasm, metabolism etc. It also helps for softening of seed coat, rupturing the seed coat, permeability of seed, for converting the insoluble food into soluble form. Water helps to bring the food and dissolved oxygen for the developing embryo.

(ii) Oxygen: Oxygen is necessary for respiration which releases the energy needed for growth. Germinating seeds respire very actively and need sufficient oxygen. Germinating seeds fulfil their O_2 requirement from the soil. Hence, for better seed germination the soil should be well drained, ploughed, aerated and well irrigated.

(iii) Suitable temperature: Moderate or optimum temperature is necessary for seed germination. The seed germinates at variety of the temperatures from 5°C to 40°C. The optimum temperature for seed germination is 20°C to 30°C.

(II) Internal Factors:

These factors are internally present in the seed like stage of embryo, mobilization of reserve food material, germination inhibitors, age and phytochrome action.

(i) Immature embryo: In some seeds the embryo is not mature. These seeds do not germinate until embryo gets matured.

(ii) Mobilization of reserve food material: During the germination embryo resume metabolic activity and undergo rapid division and expansion. The stored complex food material in the form of starch, protein or fats need to be converted into simple forms. The energy produced during respiration is used for this process.

(iii) Germination inhibitors: Some chemicals like ABA (abscisic acid) inhibits the germination. The ABA concentration increases during the dormancy of the embryo.

(iv) Age: Age of the seed is also an important factor which affect the process of germination. The early or premature seeds fail to germinate due to immature embryo and other insufficient substances. The mature or naturally aged seeds show healthy germination and can produce healthy seedling.

(v) Phytochrome action: Some seeds are sensitive to phytochrome action (PR and PFr forms). When the seeds are first exposed to Red (660 nm) and later exposed to Far-Red (730 nm) seeds will give the response or act with effect to last exposure i.e. Far Red. Red light (R) stimulates the rate of germination. If exposure to Red light (R) is followed by Far-Red light (FR) it show stimulatory effect. The important light factor for seed germination is quality of light to which the seeds are exposed. This can be shown as below:

R – Germination

R + Fr – No germination

R + Fr + R – Germination

R + Fr + R + Fr – No germination

R + Fr + R + Fr + R – Germination

The light requirement of the seeds may be replaced by hormones like Gibberellins or cytokinins.

4.7 BIOCHEMICAL CHANGES DURING SEED GERMINATION

Seed germination is a complicated process. Several changes takes place biochemically for seed germination. Seeds may be endospermic or non-endospermic mostly similar pattern of biochemical reactions is observed.

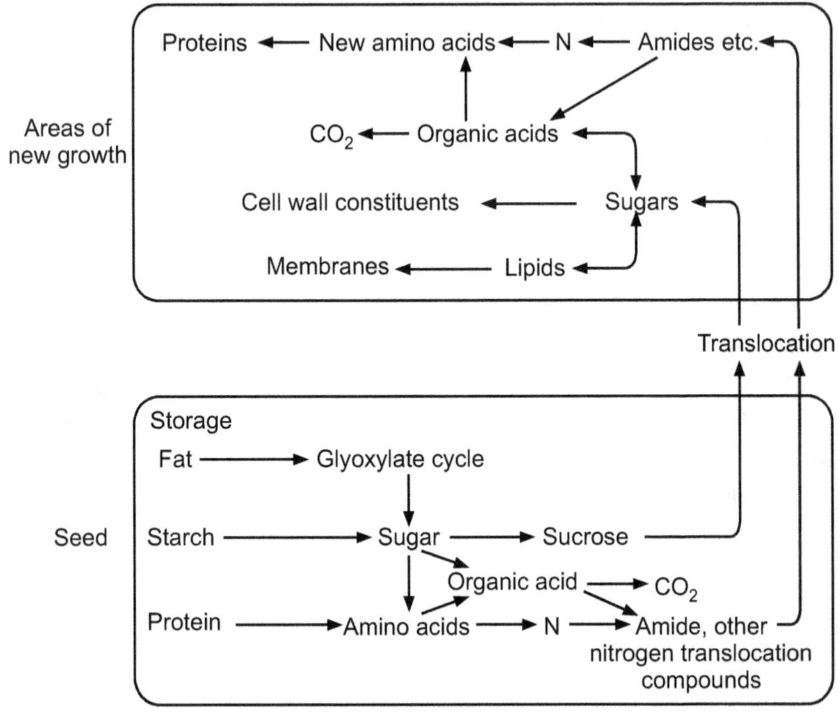

Fig. 4.5: Diagrammatic representation of biochemical changes during seed germination

During the biochemical reactions of seed germination several types of metabolites like starch, proteins, fats and other polysaccharides are hydrolysed and mobilized for the developing embryo and seedling. During the germination the hydrolytic enzymes are synthesized due to the activity of GA. In most of cereal grain, the endosperm is starchy and is surrounded by aleurone layer. Enzymes like hydrolases are secreted in this layer. The activity of β-amylase, α-amylase and enzyme protease increased during the germination. The seeds where fat is stored as reserve food, during germination it is converted into sugar and transported to growing embryo. In such seeds fats are converted to acetyl-CoA through β-oxidation pathway. Acetyl-CoA enters glyoxysomes and undergo glyoxylate cycle. In this cycle two molecules of acetyl-CoA are converted into one molecule of succinate. Succinate is converted to oxalo-acetic acid (OAA) which gives rise to phosphoenolpyruate (PEP). Through reversal of glycolysis, PEP is converted into sugars. ATP and reducing power is made available from glyoxylate cycle.

Seeds show different types of reserve food materials like fats, starch or proteins. By several pathways these are finally converted into sucrose, amides and amino acids etc. These compounds are lastly made available to the growing embryo and germinating seeds.

QUESTIONS

A. Multiple Choice Questions:

1. The seed is called dormant and the phenomenon is termed as ....
   - (a) dormancy
   - (b) scarification
   - (c) stratification
   - (d) vernalization

2. Dormant seeds are also called as resting stage .... seeds.
   - (a) viable
   - (b) non-viable
   - (c) quiescent
   - (d) non-quiescent

3. Cocklebur plant is the example of ....
   - (a) Seed coat impermeable to water.
   - (b) Seed coat impermeable to oxygen.
   - (c) Period of after ripening
   - (d) Germination inhibitors

4. Period of after ripening is completed only in the presence of moisture and low temperatures and condition called as ....
   - (a) scarification
   - (b) dormant
   - (c) stratification
   - (d) inactive seed

5. Generally in the pulp of juice of fruit the inhibitors are present in .... family plants.
   - (a) Rutaceae
   - (b) Asteraceae
   - (c) Chenopodiaceae
   - (d) Ranunculaceae

6. .... is a treatment by which the hard seed coat is ruptured or weakened.

(a) Stratification   (b) Scarification

(c) Non- scarification   (d) None of above

7. Seeds in which germination is stimulated by light are called ....

(a) Positively photoblastic

(b) Negatively photoblastic

(c) Both

(d) None of above

8. .... growth hormone break the seed dormany.

(a) Auxin   (b) Gibberellin

(c) Ascorbic acid   (d) Abscissic acid

9. The ressumption of active growth on the embryo part is known as ....

(a) Seed germination (b) dormancy

(c) viability   (d) hybrid vigour

10. The optimum temperature range for seed germination is ....

(a) 5°C to 10°C   (b) 40°C to 50°C

(c) 20°C to 30°C   (d) 50°C to 60°C

11. Entry of the water in the inner part of the seed is called as ....

(a) vernalization   (b) photoperiodism

(c) infiltration   (d) imbibition

12. Seeds which need light for germination are called as ....

(a) Positively photoblastic

(b) Negatively photoblastic

(c) Non-photoplastic

(d) pseudo-photoplastic

13. In seeds food material is stored in ………………….

    (a) embryo   (b) seed coat

    (c) epithelial layer   (d) cotyledons

14. The type of seed germination where the seed remain inside the soil is known as ………………….

    (a) hypogeal   (b) epigeal

    (c) viviparous   (d) parageal

15. The type of seed germination where the seed come out side the soil is known as ………………….

    (a) epigeal   (b) hypogeal

    (c) viviparous   (d) parageal

16. Seeds germination taking place on the mother plant is known as ………………….

    (a) viviparous   (b) hypogeal

    (c) epigeal   (d) parageal

17. …………………. phytochromes are harmful for seed germination.

    (a) Prr   (b) Pr

    (c) PFF   (d) PFr

> **Answer key:**
> (1) – a, (2) – c, (3) – b, (4) – c, (5) – a, (6) – b, (7) – a, (8) – b, (9) – a, (10) – c, (11) – d, (12) – a, (13) – d, (14) – a, (15) – a, (16) – a, (17) – d.

B. Broad Questions:

1. What is dormancy? Explain causes of dormancy.
2. What do you mean by dormancy? Explain methods of breaking seed dormancy.
3. Explain in detail process of seed germination.
4. What are different types of seed germination?
5. Enumerate the factors responsible for seed germination.
6. Explain the biochemical changes taking place during seed germination.

C. Write Short Notes on:

1. Seed dormancy.
2. Causes of seed dormancy.
3. Conditions necessary for seed germination.
4. Viviparous seed germination.
5. External factors affecting seed germination.
6. Internal factors affecting seed germination.

www.ingramcontent.com/pod-product-compliance
Lightning Source LLC
Chambersburg PA
CBHW051831150726
47998CB00001B/379